Field Guide
to the Birds of Cuba

CUBAN TROGON
Priotelus temnurus

BEE HUMMINGBIRD
Mellisuga helenae

CUBAN TODY
Todus multicolor

CUBAN PYGMY-OWL
Glaucidium siju

CUBAN GRASSQUIT
Tiaris canora

ORIENTE WARBLER
Teretistris fornsi

BLUE-HEADED QUAIL-DOVE
Starnoenas cyanocephala

FIELD GUIDE TO THE
Birds of Cuba

ORLANDO H. GARRIDO

and ARTURO KIRKCONNELL

Museo Nacional de Historia Natural de Cuba

Illustrated by Román F. Compañy

Comstock Publishing Associates

A DIVISION OF *Cornell University Press*

ITHACA, NEW YORK

First published 2000 by Cornell University Press
First printing, Cornell Paperbacks, 2000

Printed in the United States of America

Library of Congress Cataloging-in-Publication Data

Garrido, Orlando H.
    Field guide to the birds of Cuba / Orlando Garrido and Arturo Kirkconnell ;
illustrated by Román Compañy.
        p. cm.
    Includes bibliographical references (p. 239).
    ISBN 0-8014-3718-0 (cloth) — ISBN 0-8014-8631-9 (pbk.)
    1. Birds—Cuba. 2. Birds—Cuba—Identification. I. Kirkconnell, Arturo.
II. Title.
    QL688.C9 G38 2000
    598'.097291—dc21   99-088784

Cloth printing   10 9 8 7 6 5 4 3 2 1

Paperback printing   10 9 8 7 6 5 4 3 2 1

This book is dedicated to Victor L. González, for the support provided for the study of Cuba's natural history. It is also dedicated to the memory of Johannes Gundlach, and his legacy, as well as to all ornithologists, bird-watchers, and other lovers of West Indian birds.

Contents

Foreword

Publication of this excellent guide to the birds of Cuba comes at a time when there is perhaps a reasonable hope for normalization of U.S. relations with this island nation and its strongly independent, gracious, and friendly people.

Cuba is worthy of this fine effort. Its great biodiversity includes 350 species of birds (with 6 endemic genera and 21 endemic species, including the minuscule Bee Hummingbird, the beautiful Cuban Trogon, and the colorful Cuban Green Woodpecker) and is of primary importance to migrant and wintering North American birds.

The authors' experience and encyclopedic knowledge reflected here yields many points of interest about bird identification, ecology, and behavior. The illustrations go beyond the needs of a field guide and are a contribution to bird art. The book covers even rarer species and vagrants, most of them illustrated. Because transients are also illustrated and depicted in their winter plumage, North American field guides now will be unnecessary in Cuba. Cubans can be truly proud of the efforts of these Cuban authors and artist.

Searches for the Ivory-billed Woodpecker (1985–1993) that succeeded in Cuba, but were too late to save what likely were the last two or three of them, were tragic for the woodpecker. But the experience greatly heightened Cuban's interest in their wildlife and led to the establishment of some new forest reserves as well as increased protection of existing reserve areas. These efforts have benefited most of the endemic species of the island, although four or five exist in very low numbers. I only wish that the Ivory-bill were yet among these.

Orlando Garrido and Arturo Kirkconnell have worked on behalf of Cuba's fauna for many years. These talented authors, unlike most authors of Latin American guides, have a lifetime of experience in Cuban ornithology. Few such guides are by "natives." I am certain that this book will well serve the many visitors to Cuba and also enhance support for the authors' and their colleagues' conservation efforts.

LESTER L. SHORT
Lamont Curator Emeritus
Department of Ornithology
American Museum of Natural History
Research Associate, National Museums of Kenya

Preface

The diversity of Cuban birds is considerable, not only in terms of species richness, but also in terms of the number of endemics. The 354 species recorded in Cuba include 285 species that regularly occur, 66 species that only very occasionally reach here, and 3 species now deemed extinct. One hundred forty-nine species, 42 percent of the total, breed on the island. An additional 8 species are considered hypothetical.

Twenty orders and 60 families are represented. Six genera are endemic: *Cyanolimnas, Starnoenas, Xiphidiopicus, Ferminia, Teretistris,* and *Torreornis.* Living endemic species number 21, among them the charming Cuban Tody, the striking and elegant Cuban Trogon, and the smallest of all birds, the Bee Hummingbird. Two species, the Cuban Macaw and the Passenger Pigeon, were last recorded in Cuba in the nineteenth century. In addition, 57 races or subspecies have been described from the Cuban archipelago.

With this great biodiversity, the time for a Cuban field guide is long overdue. Until now, the only handbook has been James Bond's *Birds of the West Indies* (1985). A guide specifically for the birds of Cuba, such as the guides available for Jamaica, Puerto Rico, and Grand Cayman, is more than justified. Our goal was to produce a compact and portable field reference to the birds of the island. Both Cubans and bird-watchers visiting from abroad will now find it easier to identify and enjoy Cuban birds. We hope this will increase support for the conservation of Cuban birds, and that ornithologists and bird-watchers alike will appreciate this detailed description of Cuba's avifauna.

Acknowledgments

It is impossible to thank the many persons who have made this book possible. Many have provided logistical support, information, much-needed criticism, or spiritual reinforcement. To all we express our deepest gratitude. The following deserve special mention.

Our deepest and heartfelt thanks and gratitude to Victor L. González for his constant belief and support to the study of Cuban natural history as well as in the West Indies. He played a key role in the production of this book.

Our heartfelt thanks to John Faaborg, Davis W. Finch, Guy Kirwan, Douglas Macrae, Andy Mitchell, A. Townsend Peterson, Elizabeth A. H. Wallace, and James Wiley for their help, patience, and friendship. They provided many valuable comments and ideas to the draft introduction and species accounts. We gratefully acknowledge William Smith's useful comments on some illustrations. Oscar Arredondo and William Suárez provided the information on fossil birds. Manuel Iturralde-Vinent wrote the section *Origin of Cuba* and prepared the distribution maps. Onaney Muñiz, Alberto Areces, and Juan Antonio Hernández revised and made valuable comments on the botanical section. Doug Wechsler of VIREO (Academy of Natural Sciences of Philadelphia) provided photographs that were valuable in the preparation of the plates.

The RARE Center for Tropical Conservation played a crucial role in the completion of this book by awarding grants to us for trips to the United States to study specimens in various American museums. We also thank the American Museum of Natural History for grants provided to visit their collections.

We are infinitely grateful to Román Compañy, whose skill and thorough dedication resulted in such fine illustrations. We are proud to have his work accompany our text, or rather, to have our text accompany his excellent depictions of Cuban birds. Alfonso Silva Lee was patient enough to translate, revise, and fine-tune the original Spanish text, at times a seemingly never-ending task.

Words fall short for expressing our very deepest, sincere, and everlasting gratitude to a most special colleague: George E. Wallace. His great professional experience and hard work are reflected throughout the entire manuscript. He provided a wealth of valuable comments, suggestions, and criticism, and spent a great deal of time reading and editing the final draft. He also shared with us his vast field experience accumulated over his 10 years of roaming the Cuban landscape, both banding and birding. His suggestions greatly improved the final quality of the illustrations. Once again we express our heartfelt thanks for his extraordinary support and his infinite patience, and for providing us with the courage needed to complete our task. George really helped make this book possible.

I, Arturo, would like to finish by specially acknowledging my parents, Hoyt and Cristina, for all their support and love throughout the years. Two mischievous but lovely children, Arturito and Karen, gave me much love and energy. Finally, I acknowledge the most special person, without whom it would have been impossible to start, let alone finish this book: my wife Rosita, who gave me love, support and patience during my long absence, both physical and mental.

Field Guide
to the Birds of Cuba

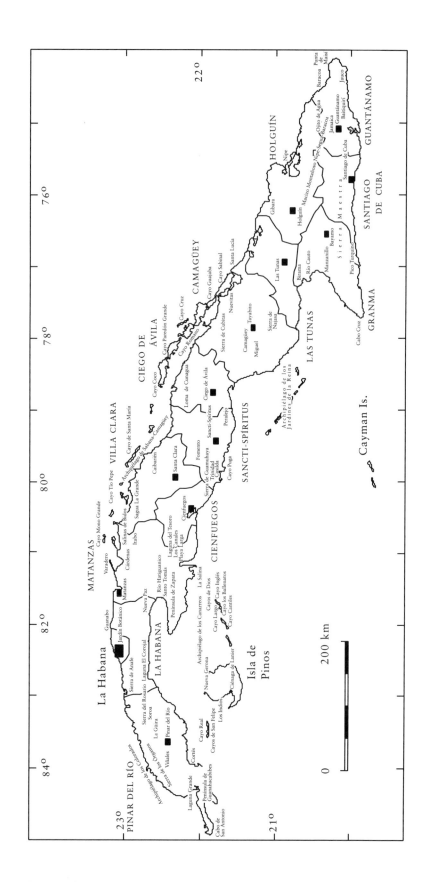

Introduction

ABOUT CUBA

With a population of about 11 million people, Cuba is the largest country in the Caribbean and accounts for over half of all the land area of the West Indies. The Cuban archipelago consists of the main island, Isla de Pinos (or Isla de la Juventud), and well over a thousand cays. Lying just south of the Tropic of Cancer, Cuba is on the interface of the Temperate and Tropical zones.

Geographic Statistics on the Cuban Archipelago

Land Area
Main island: 105,007 sq km
Isla de Pinos: 2,200 sq km
Keys: 3,715 sq km
Total area: 110,922 sq km

Geographical Limits:
Northern limit: Cayo Cruz del Padre; 23° 17' 09" north latitude
Southern limit: Punta del Inglés; 19° 49' 36" north latitude
Eastern limit: Punta del Quemado; 74° 07' 52" west longitude
Western limit: Cabo de San Antonio; 84° 57' 54" west longitude

Neighboring Countries:
Haiti: 77 km east across the Windward Passage
Jamaica: 140 km south across the Strait of Colón
United States: 180 km north across the Straits of Florida
Mexico: 210 km west across the Straits of Yucatán

Length and Width of Main Island:
Longitudinal axis: 1,250 km
Widest part: 191 km
Narrowest part: 31 km

Highest Elevations:
Western provinces: Pan de Guajaibón; 699 m, Cordillera de Guaniguanico
Central provinces: Pico de San Juan; 1,140 m, Macizo de Guamuhaya
Eastern provinces: Pico Turquino; 1,972 m, Sierra Maestra

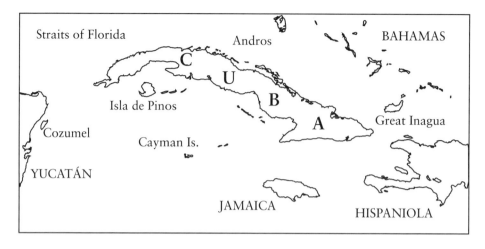

Length of Main Rivers:
Flowing north: Sagua la Grande, 144 km; Caonao, 132 km; Toa, 118 km
Flowing south: Cauto, 343 km; Zaza, 145 km; Agabama, 118 km

Climate

Cuba's climate is tropical, but varies considerably, both geographically and seasonally. The dry season usually extends from November to April. The driest regions are on the southern coast, east of Guantánamo and east of Cienfuegos, where the annual precipitation is less than 200 mm. Overall, the mean annual temperature of the island is 25.2°C. Mean annual humidity is 80 percent, and mean annual rainfall is 1,374 mm. The Sierra Maestra receives an average of 1,600 mm. The greatest annual rainfall is in the easternmost mountain range (Nipe-Sagua-Baracoa), with an average exceeding 3,400 mm. Wind direction varies by locality. Along the northern coast and inland, prevailing winds come from the north to east-northeast. Along the southern coast, winds come mostly from the northeast to southeast.

Winter weather is mediated by cold fronts from the north. These generally affect the western two-thirds of the Cuban archipelago, and occur from September through March. From February to April, southerly winds predominate.

The Caribbean experiences a marked tropical storm season from June to November. In the past 175 years, approximately 165 storms have crossed Cuba, an average of nearly one storm per year. Of these, 29 were intense, classed as hurricanes, with winds of over 210 kph. Storms are more frequent on the western third of the main island.

Origin of Cuba

Cuba has a variety of landscapes and ecosystems that are a consequence of the complex geological composition and structure of the subsoil, in combination with the tropical climate. However, in the past, the shape of the land-mass lying above sea level and the depth of the sea around the islands were different.

According to the most recent geological information regarding the origin of Cuba, the roots of the present island rose above sea level nearly 42 million years ago, only as isolated cays and shallows. These emergent cores grew large over time, and 33 to 35 million years ago became a long peninsula extending from northwestern South America to present-day central Cuba. This peninsula, dubbed *Gaarlandia* (Greater Antilles and Aves Ridge), created a pathway for some continental land animals (mammals, reptiles, amphibians, invertebrates) to colonize the Antilles. Fossil remains of these early animals have been found in Puerto Rico, including 33- to 35-million-year-old land mammals, crocodiles, and sirenians. Approximately 30 to 32 million years ago, the sea level rose and the large peninsula gave way to a group of isolated archipelagos. The connection with South America was terminated, owing to submergence of the Aves Ridge. In the Cuban region of the Caribbean, three archipelagos (eastern, central, and western) developed, separated by a channel. With temporal variations in size and relief, these archipelagos lasted until 3 or 4 million years ago. These archipelagos had vegetation similar to that of the present-day Antilles, but a different fauna. Fossil remains of these animals are found in Puerto Rico (15–20-million-year-old crocodiles, boids, and sirenians), in the Dominican Republic (15–20-million-year-old crocodiles, lizards, frogs, sirenians, and land mammals), and in Cuba (15–20-million-year-old land mammals, crocodiles, and sirenians). Fossil remains of birds have been reported from this period in Cuba (a bird bone) and the Dominican Republic (feather in amber).

Nearly 3 to 4 million years ago, the archipelagos united to form a large landmass similar in shape to the present-day main island of Cuba and Isla de Pinos. The land animals grew in number and diversity due to the enlargement of the land area, and the forests were populated by mammals (sloths, primates, rodents, insectivores, chiroptera), reptiles (lizards, crocodiles, serpents), and amphibians. Birds were abundant and occupied the niche of the main predators. Since no large carnivorous mammals existed in the islands, gigantic birds of prey (owls, hawks, etc.) evolved. The fossil remains of this unique fauna are preserved in cave deposits. The fossil record suggests that sloths, primates, and gigantic birds of prey were possibly still present in the islands at the time of the earliest settlement by humans, a mere 7,000 to 8,000 years ago.

Fossil Birds

Ancestors of present-day plants and land animals, including birds, probably arrived in Cuba beginning in the late Eocene (42–40-million years ago), when several patches of land emerged along the northern Caribbean, as mentioned already.

Studies of Cuban fossils have centered on the larger species from the late Pleistocene to Holocene. Many of the bird fossils belong to giant predators, and offer a clue to the richness of the unknown prehistoric bird life, but there is practically no information on the smaller birds which must have been the more abundant. This fact makes it difficult to trace the origin of the endemic species.

Twenty-four fossil species of birds are known from Cuba, including *Ciconia maltha* (Ciconiidae), *Teratornis* sp. (Teratornithidae), *Gymnogyps varonai*, *Sarcoramphus* sp. (Cathartidae), *Xenicibis* sp. (Threskiornithidae), *Geranoaetus melanoleucus australis*, *Titanohierax borrasi*, *Amplibuteo* sp. (Accipitridae), *Caracara creightoni*, *Milvago* sp. (Falconidae), *Grus cubensis* (Gruidae), *Nesotrochis picapicensis* (Rallidae), *Burhinus* sp. (Burhinidae), *Ara tricolor* (Psittacidae, extinct in the late 19th century), and a Nightjar (*Siphonorhis daiquiri*) with two congeners from Hispaniola and Jamaica. Evidence shows that the order Strigiformes was quite diverse in the ancestral avifauna. Within the family Tytonidae there were two species, *Tyto noeli* and *T. riveroi*, the first of which seems to have been abundant, judging by the large quantity of fossil bones. Both were larger than the Northern Eagle Owl (*Bubo bubo*) of northern Europe. Other taxa were *Otus* sp., *Bubo osvaldoi*, *Pulsatrix arredondoi*, and four species of the genera *Ornimegalonyx* (*oteroi*, *minor*, *gigas* and *acevedoi*), the latter with an estimated height of 43 inches (1.10 m). The genus *Ornimegalonyx* contains some of the world's largest known owl species. Among passeriformes, a tapaculo, *Scytalopus* sp. (Rhinocryptidae), has been found.

Derivation of Cuba's Avifauna

The Antilles and the Bahamas have a total of about 150 endemic species, most of them restricted to single islands. The 38 endemic genera present in the Antilles belong to 14 families, and among these only one, Dulidae, is itself endemic. Its single species, the Palmchat (*Dulus dominicus*), may be related to forms existing today in the southwestern United States and Mexico (genus *Phainopepla*, family Ptilogonatidae), and North America (genus *Bombycilla*, family Bombycillidae). It is difficult to understand why Dulidae is absent from Cuba, this being the largest of the Antillean islands and the one with the most diverse avifauna. A second family, once considered strictly Antillean, Todidae, also lived in North America and France, as shown by fossil remains from the Oligocene (> 24 million years ago). However, molecular studies reveal too little DNA divergence between todies and several groups of coraciiform birds (kingfishers, jacamars, motmots) for the Todidae to be of such age. Alternatively, todies may have evolved uniquely in the West Indies from a kingfisher-like colonist.

The original elements of the modern West Indian endemic bird fauna are primarily from Central America and northern South America, and arrived in the region through dispersal over the sea and possibly also through the long *Gaarlandia* peninsula (see Origin of Cuba). The island with the largest number of endemic species is Jamaica, with 27, whereas 21 species are endemic to Cuba. The avifauna of the Lesser Antilles is rather uniform, and many species exhibit a large South American influence. The Bahamas have been strongly influenced by North America. By contrast, the origins of Cuban birds are apparently more mixed, with native Cuban birds being derived mainly from North America, but with some Central American and, to a lesser degree, South American influence.

Migration

The majority, 70 percent, of Cuban bird species are migratory. Of these, 114 species are regular winter residents, and another 50 species occur only as transients. Nearly all come from North America, where they breed in summer, with only a few arriving from the southern regions to breed in Cuba.

A few migrants appear in mid-July, followed by massive arrivals beginning in August, with a peak in October. It is common to observe arriving migrants into November. The northward movement of overwintering migrants begins in late February, reaches its peak in April, and ends in mid-May. Other Nearctic migrants from south of Cuba appear in early February and are also the last migrants observed, being seen as late as June. Fall migrants arrive all along the northern coast of the island, although the central region—from Matanzas to Nuevitas—seems to receive a larger share. Many are known to winter at Guanahacabibes peninsula and Zapata peninsula, where migrants are especially abundant and diverse. Cuba is undoubtedly the most important wintering ground for Nearctic migrants in the West Indies.

Migratory summer residents, numbering some 14 species, arrive mainly from South America in late January, with a peak at the end of March. Most of these birds remain in Cuba throughout the summer and depart between September and October.

Cuban Ornithology

Bird study in Cuba began approximately 170 years ago when M. A. Vigors published a catalogue entitled *On Some Species of Birds from Cuba*, in which 45 species were reported. During 1846–48, Miguel Rodríguez Ferrer prepared a more complete catalogue, not published until 1876, that contained accounts for 207 species. A catalogue published in 1850 by Juan Lembeye contained 222 species. Truly outstanding, however, was the work accomplished by the German-born Johannes Gundlach. One of his two books, *Ornitología Cubana* (1893), describes a total of 263 species, with much valuable information on their natural history. This remains the most complete treatment of Cuban birds published to date.

More recently, during the first half of this century, both Cuban and foreign professionals have contributed a variety of publications on systematics and ecology. During the past 20 years, efforts have been directed mainly at integrated ecological studies of particular environments, bird species as such rarely being the subject of detailed and prolonged observation. As a result, after having more or less altered all habitats in the archipelago, we still have a very poor knowledge of the natural history of even the more ubiquitous endemic species. This is both a sad and a dangerous state of affairs.

Conservation

Modern conservation efforts began only as recently as 1976, when a series of reserves was identified and given different degrees of protection. These were wisely selected by a team of biologists and geographers who had in mind the need to guarantee the survival of the many unique Cuban plants and animals.

There are 80 protected areas of national relevance throughout the Cuban archipelago, with a total area of 1,338,299 ha. This includes terrestrial and marine ecosystems.

Many animal species, 37 of them birds, were declared threatened to varying degrees. At present we propose a total of 30 threatened species (see List of Threatened Species). Three of them, the Ivory-billed Woodpecker, Bachman's Warbler, and the Cuban subspecies of the Hook-billed Kite, have rarely been sighted in the past, and perhaps are already gone.

The Ivory-billed Woodpecker was last seen in 1987, and was not found again in three subsequent expeditions. Recently, though, there have been two unofficial sightings of the bird. The kite has been reported only three times in the last 30 years, the last in 1992. The warbler was last seen in Zapata peninsula in 1962 and 1964. The Zapata Wren and the Zapata Rail are very restricted in their distribution and are therefore extremely vulnerable.

List of Threatened Species

Classification (*see* Status section for definitions):

a. Critically endangered (***)
b. Endangered (**)
c. Vulnerable (*)

Pterodroma hasitata (**Black-capped Petrel**)**
Phaethon lepturus (**White-tailed Tropicbird**)**
Dendrocygna arborea (**West Indian Whistling-Duck**)*
Nomonyx dominicus (**Masked Duck**)*
Chondrohierax uncinatus (**Hook-billed Kite**)***

Accipiter striatus (**Sharp-shinned Hawk**)**
Accipiter gundlachi (**Gundlach's Hawk**)*
Cyanolimnas cerverai (**Zapata Rail**)**
Grus canadensis (**Sandhill Crane**)**
Charadrius alexandrinus (**Snowy Plover**)**
Charadrius melodus (**Piping Plover**)*
Columba inornata (**Plain Pigeon**)**
Geotrygon caniceps (**Gray-headed Quail-Dove**)*
Starnoenas cyanocephala (**Blue-headed Quail-Dove**)*
Aratinga euops (**Cuban Parakeet**)*
Amazona leucocephala (**Cuban Parrot**)*
Asio stygius (**Stygian Owl**)*
Mellisuga helenae (**Bee Hummingbird**)*
Colaptes fernandinae (**Fernandina's Flicker**)*
Campephilus principalis (**Ivory-billed Woodpecker**)***
Tyrannus cubensis (**Giant Kingbird**)**
Vireo crassirostris (**Thick-billed Vireo**)***
Corvus palmarum (**Palm Crow**)*
Tachycineta cyaneoviridis (**Bahama Swallow**)*
Ferminia cerverai (**Zapata Wren**)**
Mimus gundlachii (**Bahama Mockingbird**)*
Vermivora bachmanii (**Bachman's Warbler**)***
Coereba flaveola (**Bananaquit**)*
Tiaris bicolor (**Black-faced Grassquit**)*
Torreornis inexpectata (**Zapata Sparrow**)*

Adverse Impact on the Environment

Adverse impact on the environment takes several forms:

1. Habitat destruction. There has been extensive land alteration due to agriculture, cattle ranching, urban development, and lumber production. Logging is common, even within protected areas, to provide firewood and charcoal. Intervals between cuts usually range in the decades, but it is often too short a time to allow adequate regeneration. At present, the northern cays (Coco, Romano, Cruz, and Paredón Grande) are among the most disturbed areas, mainly as a result of development for tourism. In these cays some species occur only in small numbers and are therefore highly threatened; examples include the Bahama Mockingbird, Thick-billed Vireo, and Zapata Sparrow.

2. Hunting. Two forms may be recognized: sport hunting, purportedly within the law, and illegal poaching. For sport hunting, there are official regulations regarding species, seasons, places, and bag limits, but these are fre-

quently disregarded by both hunters and wardens. Birds are taken in numbers exceeding the legal limits and protected species, such as Ruddy Duck and Masked Duck, are sometimes taken as well. Poachers, on the other hand, persistently violate official restrictions, and use not only guns but also several kinds of traps. Throw nets, of the type used to capture herring in shallow water, have been used to capture whole flocks of Northern Bobwhite cornered by dogs. The only penalties are confiscation of prey and confiscation of the guns, which happens rarely.

3. Introduction of exotic species. Both accidental and intentional introductions plague Cuba and other islands in the Antilles. Rats, as well as feral pigs, cats, and more rarely, dogs, are now found even in the most remote and virgin forests. These animals, along with the mongoose (*Herpestes auropunctatus*), have undoubtedly had a considerable adverse effect on native bird species, although this has never been documented. Less direct, though perhaps more insidious, has been the impact on habitats of the white-tailed deer (*Odocoileus virginianus*), which has maintained wild populations in many areas for the past 150 years. In addition, over the past two or three decades, Cuba has received more than its share of exotics (African and Asian cervids and monkeys), along with more wild boars. The monkeys, about three different species, have been released on several cays off both the northern and southern coasts, with fragile ecosystems that support several unique bird and reptile races.

4. Illegal commerce. The species most affected is the Cuban Parrot. Chicks are obtained in the wild by the felling of nesting trees, and then reared by hand until fully developed. No doubt, hundreds have been smuggled out of the country in a drugged state, simply inside the pockets of travelers. An unfortunate side effect of tree-felling to obtain parrot chicks is that cavities and trees suitable for cavity excavation by other birds may become rare or completely disappearing. Such is the case for the Cuban Screech-Owl, Cuban Pygmy-Owl, Cuban Parakeet, Fernandina's Flicker, and other woodpeckers. Other species known to have been smuggled out of the country are the Cuban Grassquit, Cuban Bullfinch, and Bee Hummingbird.

5. Chemical pollution. Until the late 1980s, industries in Cuba were developing at a considerable pace. Not surprisingly, sugar refineries and electrical generating plants released a wide range of pollutants that lowered the quality of the air, soil, and water. Landscapes downstream from the larger sugar mills are usually bleak. Reports on studies of the amount and effects of these pollutants, if indeed they exist, have not been available to us.

We believe, however, that the most serious pollution has occurred through excessive use of pesticides, particularly malathion and synthetic pyrethroids. These have been used quite indiscriminately to control insect plagues in agriculture, as well as to keep mosquito populations under control, both inside cities and at tourist resorts. The impact of aerially applied pesticides on the bird populations in the cays, such as on Cayo Largo, Cayo Coco, and Cayo

Guillermo, although never properly assessed, has been great enough to produce visible and widespread avian mortality.

Bird-Watching in Cuba

Within the Caribbean, Cuba is undoubtedly the best island for bird-watching. First of all, it is home to the largest number of species, and secondly, it has several truly exceptional birding spots, such as Zapata peninsula, that will richly reward even the most demanding of bird-watchers.

Many Cubans, especially professionals and guides, speak adequate English to assist the visitor. If not, they are more than willing to communicate in whatever poor Spanish the visitor can manage.

Visiting the island during the winter has the advantage of a wider range of bird species. Both winter residents and transients will be present during this time. If a special interest in the transients exists, then the best times are from August to October and from February to April.

During the summer there is the advantage of greater activity of the native species. This is the breeding season, with much courtship activity, increased vocalization, and nest building.

Horse flies are rarely a nuisance, and then only at a few remote beaches. Mosquitoes can be a problem in the rainy season, especially a few weeks after its onset. At this time, adequate defenses must be used when walking through the woods near the coast, or in the cays, since even during the daytime mosquitoes are at times unbearable. We recommend that you wear long-sleeved shirts and long pants, and use lots of repellent. During the dry—and colder—season, most of the island is almost a bug-free paradise.

In a few areas, like Najasa and La Güira, a tiny tick locally called *garrapatilla* can be a nuisance for people sitting or laying on the ground. The itch is very annoying, but the tick can be disposed of by rubbing a drop of oil on it.

There are several mildly poisonous plants in Cuba, but if you stay on the paths, you should not have any problems. The most common of these noxious plants belongs to the genus *Comocladia*. The plant is easily recognizable by its compound, bright-green, sawlike leaves.

A comprehensive list of the major birding spots will be included in a book we are currently preparing, entitled *How, When, and Where to Find Birds in Cuba*. For present purposes, we will note only the best locations:

a. Zapata peninsula: 203 reported species; 83 percent of the endemics present, including Zapata Rail, Zapata Wren, and Zapata Sparrow.

b. Cayo Coco: over 200 species reported; 37 percent of the endemics, among them the Cuban Gnatcatcher and Oriente Warbler, and a local race of the Zapata Sparrow.

c. La Güira: 100 reported species; 50 percent of the endemics; a good place to hear the ethereal voice of the Cuban Solitaire.

d. Najasa: 100 species reported; 54 percent of the endemics; noteworthy species include Plain Pigeon, Palm Crow, and Giant Kingbird.

Where to Focus Attention on Birds

This section is for beginners or persons not familiar with some groups of common birds in Cuba. For each bird group we present the field features that an observer should consider to distinguish one species from another.

In each observation, the first thing to note is the general size and plumage color of a bird. Afterward, look for more detailed field marks, as indicated in the following list:

1. Grebes: bill and head size; white patch on wings present or absent.
2. Cormorants: bill and tail size; throat pouch color; habitat.
3. Herons and egrets: bill and leg color.
4. Ducks: bill shape and color; color of the speculum; forewing patch present or absent; underwing pattern.
5. Rails: bill size; leg color; upperparts and underparts color and pattern.
6. Gallinules and coots: bill color, streaks on sides present or absent, color pattern in undertail coverts.
7. Plovers: Upperparts color; band pattern on breast; bill size and leg color.
8. Shorebirds: bill shape and size; leg color; rump pattern; in flight, wing stripe present or absent.
9. Gulls and terns: head, bill and leg color; in flight, wing-tip pattern; shape of tail: squarish, notched, or deeply notched.
10. Hawks: underwing and tail pattern in flight.
11. Falcons: underparts and tail pattern; wing shape.
12. Pigeons and doves: color on head; pattern and color on folded wing; tail shape.
13. Quail-Doves: color and pattern on head; color on back.
14. Nightjars: color pattern on throat and breast; pattern on tail; long bristles around mouth curved or straight.
15. Flycatchers: crown flat or crested; bill size and color of lower mandible; markings around eyes; wing bars present or absent.
16. Swallows: underparts and forehead color; tail shape.
17. Thrushes: upperparts color; underparts pattern; eye ring present or absent; color of the area between the eye and the base of the bill.
18. Vireos: presence of wing bars and spectacles, or eyebrow and plain wings; color of the spectacles; color of sides and eye.
19. Warblers: Wing bars and tail spots present or absent; Back, underparts and sides streaked or not; rump color; tail wagging, flicking, or fanning.
20. Sparrow: head pattern; bill and leg color; streaked or not below; rump color; tail rounded or notched.
21. Blackbirds: bill size and shape; eye color; shape of tail.

Bird Habitats

Knowledge of Cuba's habitat types can assist greatly in discovering its birds. The vegetation of Cuba is diverse. Hundreds of different plant communities have been described for the archipelago. The native Cuban flora numbers approximately 6,350 species, 51 percent of which are endemic. Among the endemic species, 15 percent are found mainly at low elevations,

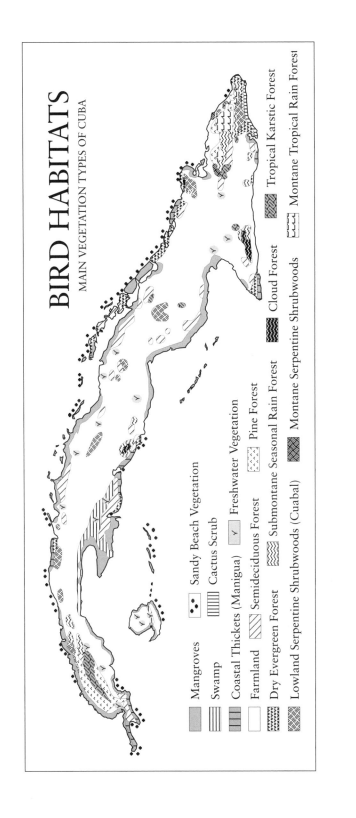

BIRD HABITATS

MAIN VEGETATION TYPES OF CUBA

Mangroves

Swamp

Coastal Thickets (Manigua)

Farmland

Dry Evergreen Forest

Lowland Serpentine Shrubwoods

Sandy Beach Vegetation

Cactus Scrub

Freshwater Vegetation

Semideciduous Forest

Submontane Seasonal Rain Forest

Montane Serpentine Shrubwoods (Cuabal)

Pine Forest

Cloud Forest

Montane Tropical Rain Forest

Tropical Karstic Forest

Montane Tropical Rain Forest

whereas approximately 75 percent are found in highland regions. One of the regions with the highest percentage of endemics is the Nipe-Sagua-Baracoa mountain system. Estimates of Cuba's forest cover range from 14 to 16 percent of the total area. Generally speaking, semideciduous forests account for about 43 percent of all forested areas. Mangroves are second, accounting for approximately 31 percent, and pine forests, relicts from past ice ages, account for about 12 percent of the forest cover.

Extensive rain forests cover the eastern mountains, whereas the southeastern coastal ranges from Guantánamo to Maisí are extremely arid. Savannas altered by agriculture are often extensive, and biologically quite distinct. Noteworthy are the sandy savannas of Isla de Pinos and Pinar del Río and those on the large stretches of flat terrain scattered between the La Habana region and the easternmost mountain ranges with their many royal palms (*Roystonea regia*). Swampy vegetation is abundant on the Península de Zapata as well as in the northeast of Ciego de Ávila province, the Ciénaga de Lanier and in the delta of the Cauto River. Zapata Peninsula is the largest wetland in the West Indies. Overall, semideciduous forests are the richest for birding, with 87 percent of all endemic species of birds occurring in them. We present a map of the main vegetation types of Cuba modified from the *Nuevo Atlas Nacional de Cuba* (Academia de Ciencias de Cuba e Instituto Cubano de Geodesia y Cartografía, 1989). The following is a summary of some important bird habitats shown on the map (the vernacular name is shown in parenthesis).

Cloud Forest

Cloud forest is found between 800 and 1,900 m in the highest reaches of the Sierra Maestra and other eastern mountain ranges, notably Pico Turquino. In this formation, principal plant species include *Cyrilla racemiflora*, and *Magnolia* spp., as well as a wide variety of ferns, mosses, and orchids. Common bird species found in this forest include the Cuban Green Woodpecker, Cuban Tody, and Oriente Warbler.

Montane Tropical Rain Forest

Montane tropical rain forest is found in areas subject to high rainfall for the majority of the year. This kind of rain forest is found in Cuchillas de Toa and Moa, at elevations up to 900 m. Principal tree species include crabwood (*Carapa guianensis*) and galba (*Calophyllum utile*), with a rich understory of ferns, orchids, and other plants. Typical bird species found here include the Cuban Solitaire, Cuban Green Woodpecker, Ivory-billed Woodpecker, Cuban Tody, and Giant Kingbird (rare).

Submontane Seasonal Rain Forest

Submontane seasonal rain forest is found in mountain ranges throughout Cuba between 200 and 800 m. It is similar to semideciduous forest (see below), but with a greater proportion of evergreen species such as *Dipholis ju-*

billa, D. salicifolia, and *Roystonea regia*. A wide variety of bird species including the Cuban Vireo, Yellow-throated Vireo, migrant warblers, Red-legged Honeycreeper, and Stripe-headed Tanager are found in this type of rain forest.

Semideciduous Forest

Semideciduous forest originally covered the majority of the island in flat and rolling regions, the lower slopes of mountains, and other seasonally humid areas. Today it occurs most notably on the Guanahacabibes peninsula, southern Isla de Pinos, central Zapata peninsula, base of the Sierra Maestra, as well as many other localities. Among the main species are cedar (*Cedrela odorata*), turpentine tree (*Bursera simaruba*), ginger tree (*Cordia gerascanthus*), hog plum (*Spondias mombin*) and lancewood (*Oxandra lanceolata*). Herbaceous ground cover is scarce. Typical birds include the Cuban Vireo, Yellow-headed Warbler, Oriente Warbler, Cuban Screech-Owl, Cuban Pygmy-Owl, Stygian Owl, Cuban Trogon, Blue-headed Quail-Dove, Gray-headed Quail-Dove and wintering migrant warblers.

Montane Serpentine Shrubwoods (Charrascal)

Found in the eastern mountains where there is a combination of high precipitation and serpentine soils, montane serpentine shrubwoods are extremely rich in endemic plants, making a brief synopsis difficult. Two of the characteristic plants are *Ilex berteroi* and *Laplacea moaensis*. The Cuban Bullfinch and Yellow-faced and Cuban Grassquits are typical birds of this habitat.

Lowland Serpentine Shrubwoods (Cuabal)

Lowland Serpentine shrubwoods comprise dry scrubby forest in low hills and flat areas over serpentine soils, and are found in small patches throughout Cuba. Common plant species include a variety of palms in the genera *Coccothrinax* and *Copernicia*. Typical birds are the Mourning Dove, West Indian Woodpecker, Northern Mockingbird, Palm Warbler (winter), and Yellow-faced Grassquit.

Pine Forest (Pinar)

Natural pine forests, containing four endemic species of pine, *Pinus caribaea, P.cubensis, P. maestrensis,* and *P. tropicalis,* are found only in Pinar del Río, on Isla de Pinos, and in the mountains of eastern Cuba. Typical species include the Olive-capped Warbler, Cuban Solitaire, Cuban Grassquit, formerly Ivory-billed Woodpecker, Red-legged Honeycreeper, and Stripe-headed Tanager.

Tropical Karstic Forest (Mogote vegetation)

Tropical karstic forest is a complex of vegetation composed mostly of thickets mainly in rendzina soils. The mogotes are limestone mounds found in Pinar del Río province and parts of central and eastern Cuba. Among the main plants are the range palm (*Gaussia princeps*), the Cayman oak (*Ekmanianthe actinophylla*), and the *protocán* (*Spathelia brittonii*). Typical species are the Cuban Solitare, Cuban Vireo, Yellow-headed Warbler, and Cuban Grassquit.

Dry Evergreen Forest

Dry evergreen forest is found mainly inland from rocky and sandy coasts. The best preserved examples of coastal forest are found in the southern part of Isla de Pinos, Guanahacabibes peninsula, and the rocky terraces of the southeastern coast. Important species are Florida poisontree (*Metopium toxiferum*), West Indian mahogany (*Swietenia mahagoni*), and turpentine tree. Typical birds are the Key West Quail-Dove, Ovenbird, and Stripe-headed Tanager.

Coastal Thickets (Manigua costera)

Distributed in the low coastal terraces of Maisí, Cabo Cruz, Cienfuegos, Guanahacabibes, and other calcareous coasts, coastal thickets are mainly composed of small bushes, such as *cuabilla de costa* (*Croton lucidus*) and several species of cactus including members of the genera *Pilosocereus* and *Dendrocereus*. Typical birds include the Northern Mockingbird, Cuban Gnatcatcher, Yellow-faced Grassquit, and Prairie Warbler. The Thick-billed Vireo and Bahama Mockingbird are found in some northern cays.

Cactus Scrub

Cactus scrub is open, thorny forest with lignum vitae (*Guaiacum officinale*), some shrubs, and abundant columnar cacti. This habitat is found in the southern coasts of Guantánamo and Santiago de Cuba provinces. Typical birds are the Cuban Gnatcatcher, Zapata Sparrow (Oriente subspecies), Northern Mockingbird, Cuban Vireo, and Oriente Warbler.

Swamp (Ciénaga)

Swamps are flooded throughout much of the year, with the vegetation growing on a layer of peat. Many extensive areas—like north of Santo Tomás—are covered with sawgrass (*Cladium jamaicensis*). The ground cover is herbaceous with areas of bushes and forests, including silver saw palmetto (*Acoelorraphe wrightii*) and cattail (*Typha domingensis*) in the wetter areas. Cyperaceous grasses and ferns are also common ground covers. Typical species include the Zapata Sparrow, Zapata Wren, Red-shouldered Blackbird, and a variety of rails and waterbirds.

Freshwater Vegetation

Freshwater vegetation is found in marshes. Some areas are extensive, such as at Laguna del Tesoro in Zapata, and some are very localized, such as in reservoirs or in ponds. Vegetation is usually floating, or rooted to the bottom. Among the common plant species are *Lemna minima, Pistia stratiotes,* and *Nymphaea* spp. Bird species using this habitat include many species of herons, ducks, gallinules and rails, and the Northern Jacana.

Mangroves (Manglar)

Mangroves are widespread along coastal plains in southern Cuba, in the northern and southern cays, and along adjacent coasts, especially from Cárdenas to Nuevitas in the north, and in Archipiélagos de los Canarreos and Jardines de la Reina in the south. Typically, there is a progression of tree species from red mangrove (*Rhizophora mangle*) at the coast, through black mangrove (*Avicennia germinans*), and white mangrove (*Laguncularia racemosa*) to button mangrove (*Conocarpus erecta*) inland. Typical birds include the Clapper Rail, Greater Antillean Grackle, Yellow Warbler, West Indian Whistling-Duck, Northern Waterthrush, Cuban Green Woodpecker, and Cuban Pewee.

Sandy Beach Vegetation

Sandy beach vegetation is typically composed of grasses, bushes, and lianas that spread along the sand, such as the beach morning glory (*Ipomoea pescaprae*). Sea grape (*Coccoloba uvifera*) is often the dominant tree species occurring in narrow swaths along the interior side. Typical birds include various migrant warblers, especially the Palm Warbler.

USING THIS GUIDE

The following are explanations of the features presented in the species accounts section of the guide.

Families

This taxonomic category is above that of genus. The family Anatidae, for example, includes the following genera: *Anas, Oxyura, Aythya,* and *Dendrocygna,* among others. These all share a set of genetic, morphological, and behavioral affinities.

The brief family description provides general information applicable to all members of this category. In parentheses at the end are the number of species in each family found in the world (represented by **W**), and the number of extant species represented in Cuba (indicated by **C**).

Species Names

We follow the systematic order, species-level changes and common and scientific names suggested by the American Ornithologists' Union (1998). The common names in English are listed first. Then the Spanish names (mostly taken from the *Catálogo de las Aves de Cuba,* Garrido and García Montaña, 1975) are presented, with the most widely used ones given first, followed by others of more restricted regional use. Many migratory species have only had names since the *Catálogo* was published. Thus, for example, most people refer to any warbler or vireo as *bijirita* or *chinchila.* Next, the scientific name of each species is given in Latin, or latinized Greek. This name is composed of two words, the first for the genus and the second for the species.

An asterisk (*) after the common English indicates that the species has been sighted at least once, but that its presence has not been confirmed by photographs or specimens.

Description

A description is given for each species, along with the most important field marks. Differences between males and females are noted when these are likely to be evident to the field observer. Color morphs are also described for species in which they are present. In the case of migratory species normally arriving in Cuba in winter plumage, this is the plumage first described. In spring, many migrant species molt and acquire a much brighter plumage, most evident in males.

The plumages of immature and juvenile birds are also described when these differ markedly from adult plumage. For some migrants, we provide descriptions of juvenile plumages even though some of these plumages have not yet been reported in Cuba. A brief description is also given of subspecies or geographical races where appropriate.

Size

The measurement given in each account is the bird's length from the tip of the bill to the end of the tail, and is given both in inches and in centimeters (in parentheses). Where size differences between males and females are large, we provide separate measurements for the sexes. When birds are over 16 inches long, fractions are discarded in favor of an approximate number. This minor imprecision should matter little to bird-watchers, for whom relative size is most important in comparing species and who rarely have the opportunity to actually measure the birds.

Behavior

Behavioral traits are excellent clues in field identification, more so when birds are distant, backlighted, or seen from an unfavorable angle. Many species behave in characteristic ways. The American Redstart, for example, constantly turns its body from one side to the other while perching, commonly with the tail fanned, and pursues and captures prey in midair. Search-

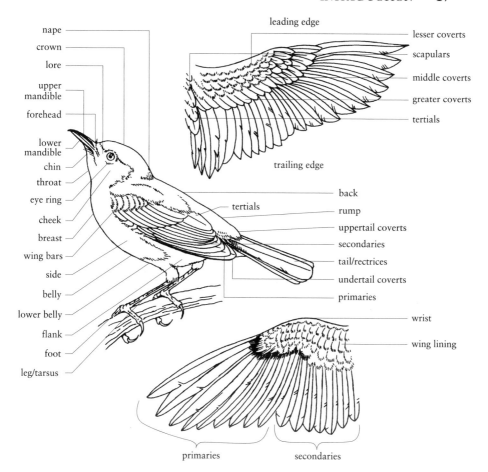

ing the undersides of leaves for insects is typical of the Northern Parula. The Northern Waterthrush characteristically moves its tail and rearbody quickly, while the Louisiana Waterthrush does the same more slowly. Knowledge of the habits of species often assists in their quick recognition.

Similar Species

Expert bird-watchers might doubt the possibility of mistaking a Cape May Warbler for a Magnolia Warbler, but to aid beginners we have included this and other similar cases under this heading. In treating many species we have listed a group of similar ones, noting the differences that are most useful for field identification. This listing starts with those most similar. Comparison with, and among, migratory species is limited to winter and immature plumages.

Range

A general description of each species' breeding distribution, both inside and outside of Cuba is given. For North American migrants, both the breeding and wintering ranges are described. In all accounts, "Cuba" refers to the main island of the Republic of Cuba.

Status

Abundant	Over 20 individuals observed per day in appropriate habitat and season
Common	Five to 19 individuals observed per day in appropriate habitat and season
Uncommon	One to 5 individuals observed in a week in the appropriate habitat and season
Rare	One to 5 sightings per year expected in the appropriate habitat and season
Very rare	One sighting every 5 years expected in the appropriate habitat and season
Vagrant	A bird that reaches Cuba by accident, or is sighted about once every 10 to 20 years
Transient	Migratory bird species that pass through Cuba, usually pausing only briefly
Summer resident	Migratory bird species that arrive from the south, principally during March, to breed in Cuba, with most departing in September
Winter resident	Migratory birds arriving in Cuba from North America in the months of July to October that spend the winter in Cuba before leaving again in March and April
Winter Visitor	Migratory birds arriving in Cuba in winter and staying no longer than a few days.

For the three categories of threat (critically endangered, endangered, and vulnerable), we follow the criteria of the IUCN Red List of Threatened Animals (1996).

In parentheses are the earliest and latest dates of most migratory species (winter and summer residents and transient species). For vagrants, the number of records, dates, and observation site are given when available.

Habitat

Under this heading are listed the types of environments where the species is most commonly found, and where we assume the birds find suitable food and cover.

Nesting

Here breeding season, nest form and materials, clutch size, and egg color are described for species that breed in Cuba. Most native Cuban land birds, including the endemics, breed mainly from March to July, with a reproductive peak in April and May. Some aquatic species breed from May to October, or even during the winter months, influenced regionally by the beginning of the rainy season and the availability of seasonal resources. Several species, for example, the Cuban Emerald, may reproduce two or three times per year.

Voice

Translating bird sounds into human vocalizations (words, sounds) is a difficult task. Nevertheless, we present a verbal description of typical vocalizations where possible. The album *Bird Songs in Cuba* (G. B. Reynard and O. H. Garrido, Cornell Laboratory of Ornithology, Ithaca, New York. 1988) contains recordings of 123 species, including most of the endemics. The album is an excellent home reference and a tape copy is useful in the field.

Food

Dietary information is provided for all species. Family descriptions usually include general information on diet.

Illustrations

Color Plates

Practically all the species so far reported for Cuba—endemics, permanent residents, migrants, introduced species, and the hypothetical records—are illustrated here. Only five species are absent from the plates. Two of these—the Common Nighthawk and Bicknell's Thrush—are practically indistinguishable from other birds of their same genera (see Species Accounts). The remaining three species not illustrated, Hermit Thrush, Black-throated Gray Warbler, and Saffron Finch, are all very recent records.

For dimorphic species, both males and females are illustrated. Juvenile or immature plumages, when different, are also depicted, in some cases for the first time in published literature (e.g., Cuban Green Woodpecker, Cuban Gnatcatcher, and Zapata Sparrow). Some endemic Cuban races are depicted for the first time as well.

Many of the migratory species are illustrated in both winter and breeding plumages. The winter plumage is emphasized because this is retained for most of their stay in Cuba, but some can be seen in breeding plumage before leaving on their northward migration.

The illustrations were drawn from specimens deposited at the Museo Nacional de Historia Natural and at the Instituto de Ecología y Sistemática. Some species absent from both collections were drawn from photographs or other sources.

Maps

These are provided for all native species. Range information is based on published literature, extensive field experience, and personal communication with colleagues. A map showing the most important localities mentioned in the species accounts is also included.

Checklist

Making a daily bird list is part of each bird-watcher's routine. These lists are of use to ornithologists for mapping the distribution of each species. Many bird-watchers consider the total number of species marked on these lists after visiting a country to be a measure of the trip's success.

The name of each species in the Index is preceded by a box, where you can place a checkmark indicating the bird has been observed.

Species Accounts

LOONS Gaviidae

Large aquatic birds with sharply pointed bill and short stiff tail. Legs are placed far back on the body, making walking very difficult. Excellent swimmers and divers, they run across the water's surface to take flight. Wingbeat is rapid, and flight is very fast, with neck and head held lower than body. Before each dive, loons make a short upward and forward movement. Food is mainly fish. They remain in the water most of the time, leaving only for breeding. Sexes alike. (**W**:5; **C**:1)

COMMON LOON
SOMORMUJO
Gavia immer   Pl. 1

DESCRIPTION. 32" (81 cm). Plumage of those reaching Cuba during winter is dark gray above, white below. Forehead steep; bill straight, stout, and grayish.

SIMILAR SPECIES. None.

RANGE. Northern half of North America, Greenland, Iceland, and Scotland. Winters along both Pacific and Atlantic coasts, south to Mexico. **Status**: Vagrant. Five records: 13 May (1971); 30 Nov, 27 Dec (1981); 23 Nov (1983); 24 Jul (1986). All along the La Habana waterfront. **Habitat**: Along coasts. **Voice**: A yodeling call. **Food**: Mainly fish, captured by diving, occasionally to depths of 50 m.

GREBES Podicipedidae

A small family of lobe-footed, entirely aquatic birds. Legs are placed far back on the body. From a distance, they look like tiny tailless ducks with slender, pointed bills. They tend to dive when approached, only rarely taking flight. Floating nests are built of aquatic vegetation. Sexes alike. (**W**:21; **C**:2)

LEAST GREBE
ZARAMAGULLÓN CHICO
Tachybaptus dominicus   Pl. 1

DESCRIPTION. 9.5" (24 cm). Easily identified in any plumage by its small size and thin straight bill, blackish in adults. Breeding adults have black

crown and throat. Large white wing patch. Eye yellow or orange. Whole bird appears dark at a distance. Nonbreeding adults have white throat. Juvenile has brown eyes, head and neck streaked. Chick has orange patch on crown, back inconspicuously streaked.

SIMILAR SPECIES. Pied-billed Grebe is larger, with a more robust bill, dark eye, wing entirely dark. Head larger, rounded.

RANGE. Mexico and Texas to northern Argentina, Bahamas, Greater Antilles. **Status**: Common permanent resident throughout Cuba, Isla de Pinos, and Cayo Coco. **Habitat**: Shallow freshwater lagoons and man-made basins, and small marshy ponds. **Nesting**: May–Jul. Builds compact nest of aquatic vegetation in the shallows, or a floating one where water is deeper. Lays four or five eggs, bluish just after laying, later spotted or stained. **Voice**: A rather long trill. Also a "ping" note that suggests the striking of a small anvil with a hammer. **Food**: Mainly aquatic insects and snails. An excellent swimmer that can remain submerged for extended periods in search of food.

PIED-BILLED GREBE
ZARAMAGULLÓN GRANDE
Podilymbus podiceps   Pl. 1

DESCRIPTION. 13" (33 cm). Brown above, including wings; paler below. A black throat patch and ring around middle of stout bill are conspicuous during breeding season. Nonbreeding adults lack black on throat and ring on bill. Juvenile has streaked head and neck. Chick has white-tipped upper mandible, orange patch on crown and nape, and conspicuously streaked blackish back. Both adults and juveniles have brown eyes.

SIMILAR SPECIES. Least Grebe is much smaller, with white wing patches, thinner bill, and yellow or orange eye.

RANGE. Southern Canada to Argentina. West Indies. **Status**: Common and widespread permanent resident in Cuba and Isla de Pinos; also reported as winter resident. **Habitat**: Fresh, brackish, and occasionally, salt water. **Nesting**: May–Jan. Compact nest is made of aquatic vegetation. Floating ones may be built over deep water. Eggs (six) are whitish and tinted with blue or buff when first laid, and somewhat larger than those of the Least Grebe. **Voice**: A long series of yelping *kaoup* notes with the tone diminishing with each burst of sound, becoming a murmur at the end, given only by the male. Both sexes also produce a churring sound. **Food**: A wide variety of aquatic animals: small fish, snails, frogs, tadpoles, insects.

PETRELS AND SHEARWATERS Procellariidae

Oceanic birds that typically fly close to the surface of the water, with rapid wingbeats alternating with stiff-winged glides. Characterized by having hooked upper and lower mandibles and a pair of large tubular nostrils over the bill. Nest is usually placed in underground tunnels or in deep cavities within the rocks on distant and isolated islands or high rocky mountain cliffs, and is visited only at night. Very rare in Cuba, found far from the coast in winter months or during storms. Sexes alike. (**W**:115; **C**:4)

BLACK-CAPPED PETREL

PAMPERO DE LAS BRUJAS; PÁJARO DE LAS TEMPESTADES

Pterodroma hasitata   Pl. 1

DESCRIPTION. 16" (41 cm). A bicolored bird with notably long wings. Black above, white below. Forehead, nape, and a wide band on base of tail and rump are white. Underwing coverts white, with a distinct black bar. Bill black. Dark-morph birds, totally black except for white rump, have never been observed in Cuba. Flight fast, with rapid wingbeats alternating with long glides.

SIMILAR SPECIES. Cory's Shearwater has grayish brown upperparts, with yellow bill.

RANGE. Caribbean, to Cape Hatteras along the Gulf Stream in the non-breeding season. Breeds on high, isolated, steep mountain sides. Extirpated from Jamaica, but breeds in Hispaniola, Guadeloupe, and Dominica. **Status**: Rare and endangered. Reported on the southern slope of Sierra Maestra mountains, near Las Brujas. **Habitat**: Oceanic. **Nesting**: In burrows or crevices on mountain cliffs. Lays one white egg.

Voice: Noisy only at breeding sites, mostly at night. Near the Sierra Maestra, the birds emit a sad, lamenting, human-like cry. **Food**: Small fish and squid.

CORY'S SHEARWATER
PAMPERO DE CORY

Calonectris diomedea   Pl. 1

DESCRIPTION. 20" (51 cm). The largest of the shearwaters occurring in Cuba. Upperparts grayish brown; underparts white. Bill thick and yellow. Flight characterized by gliding interspersed with deep slow wingbeats, generally close to the water surface. Sometimes soars.

SIMILAR SPECIES. (1) Sooty Shearwater is dark all over, except for whitish underwing coverts. (2) Black-capped Petrel is black above, with white collar and rump. (3) Audubon's Shearwater is much smaller and more crisply black and white.

RANGE. An open-water Atlantic Ocean wanderer; reaches the Mediterranean Sea and Indian Ocean. Breeds in Azores, Madeira, and Canaries and on the Mediterranean coast. **Status**: Vagrant. Three records: 26 Nov (1951); 3 May (1965); Dec (1966). All offshore at Gibara. **Habitat**: Oceanic. **Voice**: Silent away from its breeding sites. **Food**: Small schooling fish. Often follows commercial fishing vessels, and whales, to feed on offal.

SOOTY SHEARWATER
PAMPERO OSCURO; PÁJARO DE LAS TEMPESTADES

Puffinus griseus   Pl. 1

DESCRIPTION. 17" (43 cm). Dark chocolate brown all over, except for whitish underwing coverts. Bill black. Arcing flight.

SIMILAR SPECIES. Other shearwaters found in Cuba are white below.

RANGE. A wanderer of open oceanic waters. Breeds in southern latitudes: New Zealand, islands near southern tip of South America, Falkland Islands. **Status**: Vagrant. Four records: 2 Jul (1936); 14 Jul, Nov (1962); 27 Sep (1986). Offshore Matanzas and Gibara. **Habitat**: Oceanic. **Voice**: Silent away from breeding areas. **Food**: Squid, small fish, and crustaceans, caught mostly on the surface, also occasionally by diving. May approach fishing vessels for offal.

AUDUBON'S SHEARWATER
PAMPERO DE AUDUBON

Puffinus lherminieri   Pl. 1

DESCRIPTION. 12" (30 cm). Blackish above; white below. Dark-morph individuals never observed in Cuba. Wings short; tail long compared to that in other small shearwaters. Undertail coverts brownish. Underside of primaries mostly white with conspicuous broad smudgy brown margins and tip. Flight fast, with fluttering wingbeats and short glides; changes direction very quickly. Flies close to the surface of the water.

RANGE. A wanderer over tropical oceans. Breeds in some Caribbean islands. **Status**: Vagrant, not yet reported breeding. Six records: 12 May (1952); 27 Apr (1963); 27 Feb (1996). Three specimens undated. Offshore Matanzas, Gibara, Cayo Coco. **Habitat**: Oceanic. **Voice**: Noisy near nest sites, but silent at sea. **Food**: Probably flying fish and squid.

STORM-PETRELS Hydrobatidae

Small, strictly oceanic birds that, like shearwaters and petrels, only approach land for breeding. Upper mandible hooked; lower one less so. A single nostril tube. Flutter close to water, pattering over the surface with webbed feet (especially Wilson's), plucking small prey. Sometimes follow ships after offal. Found both alone and in flocks. Sexes alike. (**W**:21; **C**:3)

WILSON'S STORM-PETREL
PAMPERITO DE WILSON
Oceanites oceanicus   Pl. 1

DESCRIPTION. 7″ (18 cm). Smallest of the storm-petrels occurring in Cuban waters. Overall chocolate brown, with a conspicuous white band on rump and white patch on lower flank. Legs long; toes with yellow webs, although web color is very inconspicuous. Tail short and square, wings rounded. In flight feet show beyond tip of tail. Follows ocean-going vessels, and has the habit of hovering over the surface of the water while feeding, pattering the surface with extended feet.

SIMILAR SPECIES. Both (1) Leach's Storm-Petrel and (2) Band-rumped Storm-Petrel have forked tails, and feet not extending beyond tail tip. (3) Black Tern has pointed bill, unmarked rump.

RANGE. An oceanic bird of the Southern Hemisphere, also common in the North Atlantic. Breeds in Antarctica, islands surrounding southern tip of South America, and southern Indian Ocean. **Status**: Vagrant. Three records: 9 May, 1 Jul (1946); 6 Dec (1958). Offshore at La Habana and Gibara. **Habitat**: Oceanic. **Voice**: Silent away from the breeding areas. **Food**: Small fish, shrimp, and offal from fishing vessels.

LEACH'S STORM-PETREL
PAMPERITO DE LAS TEMPESTADES
Oceanodroma leucorhoa   Pl. 1

DESCRIPTION. 8″ (20 cm). Dark chocolate brown, with a divided white band on rump. Wings long, pointed and angled. Bill, legs, and feet (including webs) black; tail forked. Does not follow ships. Flight erratic, with sudden, sharp, directional changes; wingbeats deep. Sometimes hovers over the surface of the water while feeding, pattering the surface with extended feet.

SIMILAR SPECIES. (1) Band-rumped Storm-Petrel is slightly larger, with a more conspicuous white patch at tail base; tail slightly forked. (2) Wilson's Storm-Petrel is more delicately built; has yellow webs between toes (visible only at close range) and a square tail. Feet extend beyond tail. (3) Black Tern has pointed bill, unmarked rump.

RANGE. A cold-water oceanic bird of the Northern Hemisphere, migrating to tropical waters in the northern winter. Breeds on offshore islands. **Status**: Vagrant. One record: 25 Jul. Off Gibara. **Habitat**: Oceanic. **Voice**: Silent away from the breeding areas. **Food**: Small shrimp, fish, plankton.

BAND-RUMPED STORM-PETREL

PAMPERITO DE CASTRO

Oceanodroma castro   Pl. 1

DESCRIPTION. 8" (20 cm). Very dark chocolate brown, with a conspicuous white rump patch. Bill, legs, and feet (including webs) black; tail slightly forked. Flight erratic and level.

SIMILAR SPECIES. (1) Wilson's Storm-Petrel has a square tail and yellow webbing between toes (visible only at close range). (2) Leach's Storm-Petrel has a distinct forked tail and a white band on rump divided by a dark-gray median area. (3) Black Tern has a pointed bill, is darker above, and has an unmarked rump.

RANGE. An oceanic bird of both tropical Atlantic and Pacific oceans. Breeds on Salvage, Madeira, Cape Verde, Ascension, and Santa Elena islands in the Atlantic, and also the Galápagos and Cocos islands in the Pacific. **Status**: Vagrant. Two records: 25 Jul (1945); 6 Dec (1964). Offshore at Matanzas and Gibara. **Habitat**: Oceanic. **Voice**: Silent away from breeding areas. **Food**: Plankton.

TROPICBIRDS Phaethontidae

Like terns at first glance, but easily separated by the long whiplike central pair of tail feathers, which can be as long as, or longer than, the body. Immatures lack tail streamers, but have a dorsal pattern of narrow black bars, unlike terns. Bill tapered and pointed. Flight direct with rapid wingbeats. They catch fish by plunging from considerable heights. They sometimes settle on the water to rest, the long tail feathers then forming a graceful arc. Tropical and truly oceanic, they only approach the coast near high cliffs, during summer months, to breed. Two species visit Cuban waters, one to breed. Sexes alike. (**W**:3; **C**:2)

WHITE-TAILED TROPICBIRD
CONTRAMAESTRE; RABIJUNCO
Phaethon lepturus   Pl. 2

DESCRIPTION. 16" (41 cm), excluding tail streamers. Like a long-tailed tern. In flight, wings have conspicuous black tips and a heavy black bar on the upper surface of the inner wing. Immature has dense black bars above, including crown. Bill is yellow in immatures; reddish orange or yellow orange in adults.

SIMILAR SPECIES. Red-billed Tropicbird has a red bill, barred back, and more extensively black primaries, but lacks black bar on inner wing. Immature has thinner black bars on back and a black collar, and lacks barring on crown. Wingbeats slower.

RANGE. Circumtropical oceans, nesting on islands. Breeds at scattered localities in the West Indies. **Status**: Rare summer resident and endangered; usually seen singly or in pairs, far off the coast. Breeds at two localities in the southeast, Baitiquirí and Cabo Cruz. **Habitat**: Oceanic, coastal. **Nesting**: Mar–early summer. Nests on cliffs or caves by the sea. Lays a single pinkish egg, heavily splotched with brown. **Voice**: A rasping shrill, *cac-cac*, *cric-cric*, or *ticket-ticket*, frequently repeated in flight. **Food**: Mainly squid and fish, captured by plunging from high above the surface. Also crustaceans when near the coast.

RED-BILLED TROPICBIRD
RABIJUNCO DE PICO ROJO
Phaethon aethereus   Pl. 2

DESCRIPTION. 19" (48 cm), excluding tail streamers. Much like a tern with a very long tail. White, with black bars on back. Tip of wings conspicuously black. Bill is robust, red. Immature has fine black bars above, but white crown and black collar.

SIMILAR SPECIES. White-tailed Tropicbird has a white back, a single thick black bar over each wing, and a yellow to reddish orange bill. Immature is boldly barred above including crown, lacks collar, and has less extensive black on wing tips. Wingbeats faster.

RANGE. Tropical oceans, except central and western Pacific. Breeds on islands. Never common; usually seen singly, sometimes in pairs, well beyond
sight of land. **Status**: Vagrant. Three records: 20 Jun (1951); 16 Feb
(1982, 1988). Off La Habana and Gibara. **Habitat**: Oceanic, coastal.
Voice: A penetrating shrill: *careek* or *kek*. **Food**: Marine animals swimming
well below the surface, captured in dives from heights of 10 m or more.

BOOBIES AND GANNETS Sulidae

Large seabirds, with rather long and pointed wings, bill, and tail; throat
pouch indistinct. Overall size and bill are larger than in most gulls and terns,
and neck is longer. Found over deep, blue waters, but rarely more than a
few kilometers off the coast. All species feed on fish and squid, by plunging
from considerable heights or occasionally from level flight. Flapping is frequently interrupted with short glides. Solitary, although Brown Booby and
Northern Gannet often occur in feeding flocks. Gather in colonies during
the breeding season, laying eggs on the ground, on small barren islands.
Sexes alike. (**W**:9; **C**:4)

MASKED BOOBY
PÁJARO BOBO DE CARA AZUL
Sula dactylatra   Pl. 3

DESCRIPTION. 32″ (81 cm). White except for brownish black tail and posterior half of wings. Bill and legs orange yellow; naked skin around the bill
black. Immature has brown upperparts, with a whitish patch or white collar on upper back, white underparts, and gray legs. In flight, they show
distinct white stripes on underwing.
SIMILAR SPECIES. (1) Red-footed Booby, both white and brown morphs,
has white tail. (2) Northern Gannet is larger, with white tail. (3) Brown
Booby has dark breast, as seen from below.
RANGE. Tropical islands of all major oceans. In the West Indies, breeds in
southern Bahamas, southwest Jamaica, and the Virgin Islands. **Status**: Vagrant. Four records: 8 Mar (1948); Nov (1969); 10 Oct (1979). Another
record in 1960s. Off La Habana, Casilda, Cayo Real. **Habitat**: Oceanic,
coastal. **Voice**: Silent away from breeding areas. **Food**: Fish, squid.

BROWN BOOBY
PÁJARO BOBO PRIETO
Sula leucogaster   Pl. 3

DESCRIPTION. 28″ (71 cm). Adults dark chocolate brown, including throat
and breast; belly and underwing coverts white. Bill and legs yellow. Imma-

ture is dusky brown overall; belly and underwing coverts slightly paler. Bill bluish gray.

SIMILAR SPECIES. Both (1) Masked Booby and (2) Red-footed Booby (white morph) are all-white below, with a black and white pattern on wing.

RANGE. Pantropical seas. **Status**: Uncommon around coasts. Colonies of up to 100 pairs are established on cays off the northern and southern coasts (Cayos: Mono Grande, Puga, de Dios, and Piedras). **Habitat**: Oceanic, coastal. **Nesting**: Jan–May. The only species of booby to breed in Cuba. No nest is built; lays two bluish eggs on bare rock or sand. Shells are covered with a white chalklike material. **Voice**: A sorrowful grunt, heard only during the breeding season. **Food**: Fish, squid.

RED-FOOTED BOOBY

PÁJARO BOBO BLANCO

Sula sula   Pl. 3

DESCRIPTION. 28″ (71 cm). There are two color morphs, brown and white. White-morph birds are white with black primaries and secondaries. Brown-morph birds are brown with contrasting white tail, rump, and undertail coverts. Both morphs have bluish bill, a black band on the posterior edge of the wings, and red legs. Breeding birds have bare skin around eye blue, pink around bill base. Immature is brown all over with a blackish bill and olive to yellow legs. Brown morph has never been observed in Cuba.

SIMILAR SPECIES. Both (1) Masked Booby and (2) Brown Booby have black tails. (3) Adult Northern Gannet has black restricted to wing tips and black legs.

RANGE. Circumtropical islands. In the Antilles, breeds off Puerto Rico (Mona, Monito, Desecho, Culebra islands) and the Virgin islands. **Status**: Vagrant. Five records: one last century (1870); subsequently 5 Nov (1933); 12 Oct (1952); Aug (1972); 14 Jan (1988). Western Cuba: Guanabo beach, Playa Larga, Cabo de San Antonio. **Habitat**: Oceanic, coastal. **Voice**: A prolonged cackling. **Food**: Fish, squid.

NORTHERN GANNET*

PÁJARO BOBO DEL NORTE

Morus bassanus   Pl. 51

DESCRIPTION. 37" (94 cm). White with mustard-color wash on head and neck; black wing tips and legs; gray bill. Immature mostly dusky, with white belly.

SIMILAR SPECIES. (1) Red-footed Booby has entire trailing edge of wing black; legs red. (2) Masked Booby also has entire trailing edge of wing black; as well as black tail and yellow legs.

RANGE. Breeds on North Atlantic islands. American populations winter south to Gulf of Mexico. **Status**: Vagrant. One record: 28 Jan (1994). Off La Habana. **Habitat**: Oceanic. **Voice**: Silent away from breeding areas. **Food**: Fish, squid.

PELICANS Pelecanidae

Among the most conspicuous aquatic birds, owing to their large size and unusual bill morphology. Silhouette is unmistakable: a heavy body with a very long bill, usually resting on neck, even in flight. Large extensible throat pouch is used for fishing, and otherwise not evident. Wings long; tail short; legs thick, with completely webbed toes. They float buoyantly. Sexes alike. (**W**:9; **C**:2)

AMERICAN WHITE PELICAN

ALCATRAZ BLANCO

Pelecanus erythrorhynchos   Pl. 2

DESCRIPTION. 60" (152 cm). A huge bird, totally white, except for the primaries and part of the secondaries, which are black and conspicuous in flight. Bill and legs are orangish yellow. Never plunges for food; fish are scooped up while swimming. Immature has mottled wing coverts; bill and legs duller. Sometimes soars.

SIMILAR SPECIES. Brown Pelican is smaller, grayish brown instead of white, and feeds mainly by plunging from considerable heights.

RANGE. Breeds in western and central North America. Winters along southern coasts of United States and Central America to Guatemala. **Status**: Vagrant. Eight records: one last century (1838); recent records include 25 Jan (1940); 25 Jan, 11 Apr (1954); Feb, 1, 5 Mar (1989); 17 Feb (1997). Yumurí and Zapata peninsula. **Habitat**: Ponds, lakes. **Voice**: Silent outside the breeding season. **Food**: Schooling fish, herded while swimming in shallow water.

BROWN PELICAN
PELÍCANO; ALCATRAZ
Pelecanus occidentalis   Pl. 2

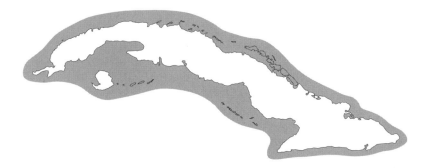

DESCRIPTION. 50" (127 cm). Large and silvery grayish brown, with head and part of neck white to yellow in adults. Breeding plumage has dark chestnut hindneck. Juvenile has dark brown head and neck. An imposing bird in flight, often seen in small flocks low over the water, forming lines diagonal to forward movement. Flapping and gliding alternately, with wing tips occasionally touching the surface.

SIMILAR SPECIES. American White Pelican is larger, mostly white, and never plunges for food.

RANGE. Coasts of southeastern and western United States to Guianas and northern Chile; southern Bahamas and Greater Antilles. **Status**: Common permanent and winter resident; also transient in Cuba, Isla de Pinos, and several cays. **Habitat**: Beaches, bays and estuaries. **Nesting**: Mar–Sep. Breeds in numerous colonies around the island, building crude nests on mangroves and laying two or three white eggs. **Voice**: Adults usually silent; at times produce a weak croaking. Nestlings squeal. **Food**: Feeds mainly on large schooling fish by plunging into water.

CORMORANTS Phalacrocoracidae

Large aquatic birds similar to ducks, uniformly black or dark brown. Bill slender; upper mandible has hooked tip. Bare facial skin and throat pouch brightly colored. Lack external nostrils. Legs set far back on body; toes completely webbed. They perch upright, with neck curved into an S shape, frequently with wings stretched out to dry in the sun. They feed on a variety of fish obtained by diving from the surface and swimming underwater. Large groups congregate for breeding or roosting, sometimes for feeding. Flocks fly low, in a straight line, with fast wingbeats. Sexes alike. (**W**:38; **C**:2)

DOUBLE-CRESTED CORMORANT
CORÚA DE MAR; CORÚA GRANDE

Phalacrocorax auritus   Pl. 3

DESCRIPTION. 32″ (81 cm). Black overall, with a green metallic sheen evident in good light. Bare facial skin and rounded orange throat pouch. Double crest is inconspicuous. Young similar to adults, but browner above and much paler below, sometimes nearly whitish on the throat and upper breast; throat pouch yellowish orange. In flight, kinked neck with head held above level of posterior portion of neck is conspicuous.

SIMILAR SPECIES. (1) Neotropic Cormorant is smaller, with smaller yellow throat pouch, tapering to point behind bill, and longer tail. (2) Anhinga is slimmer, with a straight, sharp-pointed bill, longer neck, and long, fan-shaped, white-tipped tail. In flight, alternates flapping and gliding, unlike more or less steady flapping of cormorants.

RANGE. Southern Alaska to Atlantic Canada, south to Baja California, Gulf of Mexico, northern Bahamas, and Cuba. **Status**: Common permanent and rare winter resident in Cuba, Isla de Pinos, and several cays. **Habitat**: Shallow inshore waters and bays, especially where bottom is muddy or covered with turtle grass. **Nesting**: Jan–Nov. Colonies breed in tall mangroves, building crude stick nests. Lays three to five, pale bluish-green eggs covered with a chalky white material. **Voice**: Generally silent, but at colonies some cackling and other guttural sounds can be heard. **Food**: Schooling fish.

NEOTROPIC CORMORANT
CORÚA DE AGUA DULCE; CORÚA CHICA

Phalacrocorax brasilianus   Pl. 3

DESCRIPTION. 26″ (66 cm). Black and long-tailed. Naked facial skin and throat pouch yellowish orange and tapering to point behind bill. In breeding plumage, the throat pouch is bordered with white and some white feathers also appear on sides of head. Young birds similar to adults, but slightly paler, especially below.

SIMILAR SPECIES. (1) Double-crested Cormorant is larger, has proportionately shorter tail, and rounded throat pouch. (2) Anhinga has a straight, sharp-pointed bill, longer neck, and long, fan-shaped, white-tipped tail. In flight, alternates flapping and gliding, unlike more or less steady flapping of cormorants.

RANGE. Southern United States to Tierra del Fuego, including some Andean lakes, Bahamas, and Cuba. **Status**: Common permanent resident in Cuba, Isla de Pinos, and some northern cays, especially abundant in Zapata peninsula and reservoirs in Camagüey province. **Habitat**: Coastal waters, rivers, large lagoons, reservoirs, and swamps bordering the sea. **Nesting**: Apr–Sep. Breeds in colonies, usually smaller than those of Double-crested Cormorant. However, at some places, such as Presa Muñoz in Camagüey province, there are large colonies of more than 1,000 pairs. Nests usually in trees or bushes growing in water. Nest is a rough cup or bulkier structure of small sticks. Lays usually four pale-blue eggs, with uneven chalky outer layer, becoming stained. **Voice**: A guttural piglike grunt. **Food**: Fish, frogs, insects.

DARTERS Anhingidae

Strictly aquatic birds, with small head, long neck and tail, and long pointed bill used to spear prey, usually fishes. Dark and slim, they resemble cormorants. More common in freshwater, where they swim and dive admirably, often swimming with only the neck and head protruding from the water. Flies with neck extended and tail fanned and with alternate flapping and gliding, occasionally soaring high in huge circles. Sexes different. (**W**:4; **C**:1)

ANHINGA
MARBELLA; CORÚA REAL
Anhinga anhinga   Pl. 3

DESCRIPTION. 35" (89 cm). *Male*: blackish brown with green iridescence. Large silvery white areas on upper wing. In breeding plumage, head and neck also become marked with white. Bill yellow; tail long with white tip.

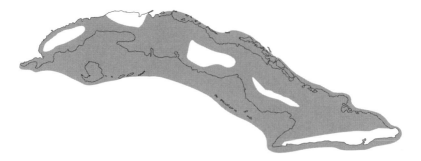

Female: Head, neck, and chest pale brown. Immature is similar to female, but paler. Spread-wing posture is typical.

SIMILAR SPECIES. Both species of cormorants have shorter tail and neck, and hooked bills. In flight their flapping is steady, practically without gliding.

RANGE. Southeastern United States through Central America to Argentina; Grenada and Cuba. **Status**: Common permanent resident in Cuba, Isla de Pinos, and several cays. **Habitat**: Estuaries and shallow lakes with calm waters, fresh or brackish to salt; also along coasts. **Nesting**: Mar–Oct. Breeds in colonies, like cormorants. Nests in trees or bushes. Nests tend to be small, built of twigs, lined with leafy twigs. Lays two to three bluish-white eggs, thinly covered with a white, chalklike material. **Voice**: Often produces harsh guttural sounds. **Food**: Mainly fish.

FRIGATEBIRDS Fregatidae

Seabirds with exceptionally long, narrow wings, very short legs, and long hooked bills. Tail is long and deeply forked, usually folded in a point. Body weight disproportionately low. Coastal, rarely straying very far offshore. In their high, motionless soaring flight, they present a majestic and unmistakable silhouette. Food, mainly fish, is captured by dramatically fast dives, swooping over the surface, with only the bill touching the water. Also feeds by harassing other seabirds in midair until food is regurgitated. Extraordinary vision allows them to congregate on an abundant food source in a matter of minutes. During the breeding season, males display by inflating a red throat pouch. Sexes different. (**W**:5; **C**:1)

MAGNIFICENT FRIGATEBIRD

RABIHORCADO; FRAGATA; GUINCHO

Fregata magnificens   Pl. 2

DESCRIPTION. 40″ (102 cm). *Male*: Entirely black, with iridescent mauve back. Throat pouch red to orange, normally invisible except when inflated. *Female*: Blackish brown, with white patch on chest. Immature has white head, neck, and breast.

SIMILAR SPECIES. None.

RANGE. Tropical America, both Atlantic and Pacific oceans. **Status**: Common permanent resident along coasts. **Habitat**: Coastal. **Nesting**: Mar–Jan. Breeds in colonies only in cays. Nest very crude, low on mangrove branches or occasionally on the ground. Lays a single white egg. **Voice**: A grunting murmur produced only during the breeding season. **Food**: Fish, jellyfish, crustaceans.

BITTERNS AND HERONS Ardeidae

Semiaquatic birds that wade in shallow water. Legs, neck, and bill are quite long; toes long and slender, outer and middle ones connected by a small web. Wings wide and rounded; tail very short. All generally fly with neck folded; thus head appears drawn back over shoulder. Some species, particularly in breeding plumage, have long plumes on head, neck, or back. A few are dichromatic. With few exceptions, they are colonial breeders that often form large mixed-species colonies. Most feed on a large variety of aquatic organisms, such as fish, frogs, crustaceans, and insects. Terrestrial vertebrates are also taken. Flight strong, with regular wingbeats. As a rule they lay pale greenish-blue eggs. Sexes alike except Least Bittern. (**W**:65; **C**:12)

AMERICAN BITTERN
GUANABÁ ROJO; AVE TORO

Botaurus lentiginosus   Pl. 4

DESCRIPTION. 28″ (71 cm). A chunky heron, rich brown, with streaked underparts and a black stripe on side of neck. Bill yellow with blackish ridge. Legs and feet greenish yellow. Dark primaries and secondaries contrast strongly with lighter-brown body and rest of wing when bird is seen in flight. Juvenile similar to adult, but without black stripe on neck. When alarmed, and "freezing" with bill pointing up, bitterns become almost invisible.

SIMILAR SPECIES. Immatures of both (1) Black-crowned Night-Heron and (2) Yellow-crowned Night-Heron are grayer, less warmly toned, with shorter bills, and lack black stripe on neck; wings show less contrast between flight feathers and coverts.

RANGE. Breeds in North America and Mexico; winters from southern United States to Panama and Greater Antilles, Swan and Cayman Islands, Bahamas, Cuba, and Bermuda. **Status**: Rare winter resident and transient (17 Aug–Apr). **Habitat**: Savannas, swamps, reservoirs with reeds. **Voice**: When rising, a repeated, grunting *coc*. A deep booming is given on the breeding grounds. **Food**: Small vertebrates, aquatic invertebrates.

LEAST BITTERN
GARCITA; MARTINETE
Ixobrychus exilis

Pl. 4

DESCRIPTION. 13" (33 cm). A very small heron; rufous, with a large buffy patch on upper wing. Male has black crown and back; female has brown back. Juvenile resembles female, with prominent streaking on back and chest. Secretive and difficult to spot, remaining hidden in reeds and flying only occasionally.

SIMILAR SPECIES. Green Heron is larger, lacks buffy wing patch.

RANGE. Eastern and southwestern United States to northern Argentina; Greater Antilles. **Status**: Common permanent and winter resident; also transient in Cuba and Isla de Pinos. **Habitat**: Pastures, swamps, lagoons, and marshes with high grass or reeds. **Nesting**: Apr–Sep. A solitary breeder; builds small, fragile nest among reeds. Lays four or five pale greenish-blue eggs. **Voice**: A series of four or five low cooing notes. Also a cackling when flushed. **Food**: Small fish and insects.

GREAT BLUE HERON
GARCILOTE; GARCILOTE CENICIENTO
Ardea herodias

Pl. 4

DESCRIPTION. 48" (122 cm). The largest of Cuban herons. Dichromatic; white morph is restricted mostly to the cays where it is locally common.

Dark morph is bluish gray, with broad black stripes on head. Foreneck streaked with black. Legs brownish green. When breeding, bill is yellow and lower neck and back are draped with long plumes. Young similar to nonbreeding adult, with black crown. White-morph birds have yellow or pinkish legs. Flies with slow wingbeats.

SIMILAR SPECIES. (1) Great Egret could be confused with white-morph birds, but has black legs. (2) Sandhill Crane has black bill, red crown, and flies with neck extended.

RANGE. Southeastern Alaska, southern Canada to Mexico, Greater Antilles, and northern South America. **Status**: Common permanent and winter resident; also transient in Cuba, Isla de Pinos, and many cays. **Habitat**: Shores, tidal flats, marshes, flooded fields. **Nesting**: Feb–Jul. Nests colonially, in trees or bushes, or ledges of cliffs. Nest is large, flat, of interwoven sticks. Lays two or three light bluish-green eggs. **Voice**: A deep, harsh, somewhat gooselike croak. Also a guttural grunt. **Food**: Mostly fish, but also nestlings, reptiles, amphibians, small mammals.

GREAT EGRET
GARZÓN; GARZÓN BLANCO
Ardea alba   Pl. 5

DESCRIPTION. 40" (102 cm). Entirely white. In breeding season, plumes reach beyond tail. Bill yellow or orange; legs black. Immatures and nonbreeding adults have duller-colored bill and legs. A patient and precise predator, often standing motionless for many minutes, but rarely missing its prey.

SIMILAR SPECIES. White-morph Great Blue Heron has yellow or pinkish legs and heavier bill.

RANGE. Cosmopolitan in warmer and temperate regions. **Status:** Common permanent and winter resident; also transient in Cuba, Isla de Pinos, and many cays. **Habitat:** Marshes, lagoons, mangroves, tidal flats, along streams. **Nesting:** Jan–Jul. Builds a large flat stick platform or bulkier reused structure, similar to, but thinner and frailer than that built by Great Blue Heron. Lays three or four pale greenish-blue eggs. **Voice:** A low hoarse croak. **Food:** Insects, aquatic invertebrates, small vertebrates.

SNOWY EGRET

GARZA BLANCA; GARZA REAL

Egretta thula   Pl. 5

DESCRIPTION. 24" (61 cm). Entirely white with black bill and legs, and yellow toes. While breeding, long plumes are present on head, neck and back; lores yellow to deep red. Immatures have back of legs pale green; some have a pale lower mandible as well and almost entirely olive legs. The most active of all herons when in search of food, constantly stirring bottom sediments by vibrating its feet, and running or jumping to seize prey.

SIMILAR SPECIES. (1) Immature Little Blue Heron has green legs and bicolored bill. (2) White-morph Reddish Egret has longer and thicker bicolored bill, thick neck, and dark lores. (3) Immature and nonbreeding Cattle Egret has shorter and thicker yellow bill and greenish legs and feet.

RANGE. Southern Canada to central Argentina; Greater Antilles. **Status:** Common permanent and winter resident, and transient in Cuba, Isla de Pinos, and several cays. **Habitat:** Fresh, brackish, and salt waters. **Nesting:** May–Oct. Nests in trees, often mangroves. Builds a shallow structure, elliptical and rather flat, of thin twigs. Lays three greenish-blue eggs. **Voice:** Generally silent; may produce a low croak. **Food:** Fish, invertebrates.

LITTLE BLUE HERON
GARZA AZUL; GARZA COMÚN

Egretta caerulea   Pl. 5

DESCRIPTION. 24″ (61 cm). Slate blue, with violaceous brown head and neck. Bill grayish blue with a black tip; lores gray. Legs green. Immatures entirely white, except for variable amount of slate on wing tips. Molting birds are white with irregular patches of slate blue. Not especially active while feeding and usually seen standing still. Often flies with neck extended.

SIMILAR SPECIES. (1) White-morph Reddish Egret has blue legs and a more contrastingly bicolored bill; dark-morph Reddish Egret has a rusty head and neck and a pink, black-tipped bill. (2) Snowy Egret has black bill and legs, with yellow feet, and (3) Tricolored Heron has white belly.

RANGE. Eastern United States, southern California to Peru, central Brazil and Uruguay; West Indies. **Status**: Common permanent and winter resident, and transient in Cuba, Isla de Pinos, and many cays. **Habitat**: Swamps, both freshwater and salt water; also lagoons, ponds, flooded savannas, quiet creeks. **Nesting**: Oct–Aug. Nests in trees, usually low down, or in shrubs. Nests colonially, often with other species where it tends to be on the outskirts of the colony. Lays three or four pale greenish-blue eggs. **Voice**: When alarmed, a harsh cackle. **Food**: Mostly fish, amphibians, aquatic invertebrates.

TRICOLORED HERON
GARZA DE VIENTRE BLANCO; GARZA MORADA

Egretta tricolor   Pl. 4

DESCRIPTION. 26″ (66 cm). Bluish gray, with white belly, white wing linings, and narrow frontal stripe along neck. Rump white. Bill mostly yellow, diffusely tipped with black. More slender necked and long billed than other small herons. Immature has chestnut hindneck and wing coverts. Quite active when fishing.

SIMILAR SPECIES. (1) Little Blue Heron and (2) Reddish Egret have dark bellies, entirely dark necks, and more strikingly bicolored bills.

RANGE. Southern and eastern United States to Peru and Brazil, Bahamas, and Greater Antilles. **Status**: Common permanent and winter resident; also transient in Cuba, Isla de Pinos, and many cays. **Habitat**: Large coastal swamps, marshes, and rocky coasts. **Nesting**: Apr–Sep. Nests in mangrove. Highly sociable, nesting in colonies. Builds a shallow structure of slender twigs, round or oval, with a hollow several inches deep. Lays three or four greenish-blue eggs. **Voice**: Various harsh guttural sounds. **Food**: Fish, other small vertebrates, aquatic invertebrates.

REDDISH EGRET

GARZA ROJA; GARZA MORADA

Egretta rufescens   Pl. 4

DESCRIPTION. 30" (76 cm). Two color morphs. Dark birds are slate gray, with head and neck purplish to reddish chestnut. White birds are entirely white. The two morphs are equally common. Bill heavy and long, bicolored in both morphs, pink with black tip. Legs and lores slate blue, both becoming blue in high breeding. Plumes on head and neck give a shaggy appearance. Dark morph immature is gray, with reddish brown head and neck. Immatures of both morphs have dark bill, lores, and legs. Rare mixed-morph birds also occur. Feeds by lurching about, extending wings to form a canopy.

SIMILAR SPECIES. (1) Little Blue Heron is darker, with smaller bill, pale blue at base. (2) Immature Little Blue Heron has green legs, and (3) Adult Snowy Egret has black bill and legs, black, with yellow feet.

RANGE. Southern United States, Mexico, northwestern Bahamas, Cuba, and Hispaniola. **Status**: Common permanent and winter resident, and transient in Cuba, Isla de Pinos, and many cays. **Habitat**: Marshes, bays, and beaches, mainly in saltwater or brackish water. **Nesting**: Apr–Jan. Nests in mangrove, at heights of up to 5 m, or on the ground. Builds a platform of thin sticks, stems, and roots. Lays three or four greenish-blue eggs. **Voice**: A guttural cackle, less harsh than that of other herons and egrets. **Food**: Small vertebrates and aquatic invertebrates.

CATTLE EGRET

GARZA GANADERA; GARZA BUEYERA

Bubulcus ibis   Pl. 5

DESCRIPTION. 22" (56 cm). Entirely white with yellow to orange bill and yellow to green legs. In breeding condition, has buff plumes on upper head, base of foreneck, and back; bill and legs become dusky red. Juvenile may have black bill. Immature lacks plumes, and resembles nonbreeding adults with green to black legs. They fly long distances to their colonies and roosts, usually in tall trees near water.

SIMILAR SPECIES. Other white herons are larger, with longer bills and necks, and are largely aquatic. Cattle Egret is the only small white heron with a yellow bill.

RANGE. Originally southern Eurasia and Africa. Invaded South America in the late 1870s and now ranges from southern Canada and United States to Chile and Argentina, including West Indies. **Status**: Abundant permanent and winter resident; also transient in Cuba, Isla de Pinos, and some cays. **Habitat**: Marshes, swamps, and farms, wherever soil is being tilled or cattle are grazing. **Nesting**: Apr–Jul. Nests in colonies, often in large trees or bushes. Builds a shallow structure of twigs, sticks, dead reeds, or other vegetation. Lays four or five bluish-white eggs. **Voice**: Usually silent, although may produce grunting sounds. **Food**: Insects, mollusks, nestlings, other small vertebrates.

GREEN HERON
AGUAITACAIMÁN; CAGALECHE
Butorides virescens   Pl. 5

DESCRIPTION. 18″ (46 cm). Blue gray, with iridescent green back. Neck dark chestnut with narrow black-bordered white stripe down foreneck. Short greenish yellow to orange legs and normally retracted neck give it a compact look. Raises a blackish, iridescent crest when alarmed. Immature is duller with streaked underparts.

SIMILAR SPECIES. Least Bittern is smaller and very secretive, hiding deep in marshes and grassy fields. In flight, the buffy ochre patch on inner wing is conspicuous.

RANGE. Southern Canada to Panama; West Indies. **Status**: Common permanent and winter resident; also transient in Cuba, Isla de Pinos, and several cays. **Habitat**: Mainly freshwater areas of all kinds, but also brackish and salty waters. **Nesting**: Mar–Aug. A solitary breeder. Nests in trees, mangroves, and thickets. Builds a small flimsy platform of twigs, reeds, and vines. Also reuses nests in successive seasons, adding material to form a bulky structure. Lays three or four greenish-blue eggs. **Voice**: A noisy bird producing a low cackling and a sudden squeal when alarmed. **Food**: Fish, insects, terrestrial vertebrates.

BLACK-CROWNED NIGHT-HERON
GUANABÁ DE LA FLORIDA; GUANABÁ LOMINEGRO
Nycticorax nycticorax   Pl. 4

DESCRIPTION. 25" (64 cm). A stocky gray and white heron with black back and crown, the latter with a few long white plumes. Neck, legs, and bill are short. Legs greenish yellow; eye red. Immature is streaked below, and grayish brown above spotted with white, with green legs and dusky yellow bill. In flight, only the feet project beyond tail. Active at night.

SIMILAR SPECIES. (1) Yellow-crowned Night-Heron is gray. Immature has less distinct white spots on back, thinner neck, shorter and thicker bill, pale only at base. Legs are longer so that feet and part of legs project beyond tail in flight. (2) American Bittern resembles immature Black-crowned Night-Heron, but is more warmly toned, has longer and thinner bill, and a black stripe on side of neck.

RANGE. One of the most widely distributed bird species in the world, absent only from Australia and polar regions. Bahamas and Greater Antilles. **Status**: Common permanent and winter resident; also transient in Cuba, Isla de Pinos, and some cays. **Habitat**: Estuaries, swamps, mangroves, rice fields, mostly in freshwater or brackish water. **Nesting**: Mar–Jul. Forms small colonies in mangroves and all types of trees. Builds a platform with a shallow hollow, of twigs and reeds. Lays four or five greenish-blue eggs. **Voice**: In flight, a characteristic *quock* often heard at dusk or at night, deeper than that of Yellow-crowned Night-Heron. When breeding, guttural sounds and penetrating croaks. **Food**: Crustaceans, fish, aquatic insects, frogs, mice.

YELLOW-CROWNED NIGHT-HERON

GUANABÁ REAL

Nyctanassa violacea Pl. 4

DESCRIPTION. 24" (61 cm). A stocky gray heron, with head boldly striped black and white; bill mostly black. Legs yellow and eye red. Immature dusky brown, with small white spots on back and narrow streaks below. In flight, feet and lower portion of legs extend beyond tail.

SIMILAR SPECIES. (1) Black-crowned Night-Heron has a solid black crown, and black back. Immature has bolder white spots on back, longer bill (lower mandible may be entirely pale), and shorter legs. (2) American Bittern resembles immature, but is a much richer brown, and has longer and thinner bill, black stripe on side of neck, and a bolder streaked pattern below.

RANGE. Southwestern and eastern United States to Peru and Brazil; Galápagos and West Indies. **Status**: Common permanent and winter resident; also transient in Cuba, Isla de Pinos, and certain cays. **Habitat**: Coasts and estuaries, in freshwater to salt water. Also in wooded areas near pools, lakes, and reservoirs. **Nesting**: Apr–Aug. Solitary or small loose colonies. Nests in mangroves, other trees, and bushes at varying heights from near ground level to 17m. Lays three or four pale greenish-blue eggs. **Voice**: A short, somewhat muffled *quack*, higher in pitch than corresponding call of Black-crowned Night-Heron. **Food**: Crabs, fish, insects, young birds.

IBISES AND SPOONBILLS Threskiornithidae

Medium-sized to large wading birds with long legs and necks. Bill is long and thin, decurved or spatulate. Ibises are social birds that fly in organized, V-shaped flocks or lines, with necks outstretched. Roosts and nesting colonies are in quiet, secluded swamps, often with herons. They feed mainly on small crustaceans and other invertebrates, obtained by probing deeply in bottom sediments while walking in shallow water. Sexes alike. (**W**:34; **C**:4)

WHITE IBIS
COCO BLANCO
Eudocimus albus   Pl. 6

DESCRIPTION. 25" (64 cm). White with black wing tips. The long decurved bill, face, and legs are red. Immature brown above, with white belly and rump. In flight, flapping alternates with gliding. Occasionally soars.

SIMILAR SPECIES. (1) Glossy Ibis resembles immature White Ibis but is all dark. (2) Wood Stork is much larger, with heavy dark bill and a broad black trailing edge to wing. (3) Immature Roseate Spoonbill has a spatulate yellowish bill and lacks black wing tips.

RANGE. Southeastern and southwestern United States to Peru, Venezuela, and French Guiana; Greater Antilles (except Puerto Rico). **Status**: Common permanent resident in Cuba, Isla de Pinos, and many of the cays. **Habitat**: Muddy coastal shores, swamps, flooded fields, brackish lagoons. **Nesting**: Apr–Sep. Nests in colonies in trees. Builds a platform of dead sticks and fresh twigs, with leaves. Usually some green leaves in the lining. Lays three large greenish-white eggs, splotched dark brown. **Voice**: Nasal growls: *oohh-oohh*, sometimes in series. **Food**: Insects, crabs, crayfish, snails, shrimps.

SCARLET IBIS
COCO ROJO
Eudocimus ruber   Pl. 6

DESCRIPTION. 23" (58 cm). Brilliant scarlet with black wing tips.
SIMILAR SPECIES. Roseate Spoonbill is paler with white neck and gray spatulate bill, and lacks black wing tips.
RANGE. South America from Venezuela to Brazil, mostly near coast. **Status**: Very rare visitor to northern coasts and cays. **Habitat**: Mostly near coast. **Voice**: Usually silent in nonbreeding season. **Food**: Insects, aquatic invertebrates.

GLOSSY IBIS
COCO PRIETO
Plegadis falcinellus   Pl. 6

DESCRIPTION. 23" (58 cm). Plumage appears black from a distance, but at close range has metallic green on wings and back. Bill long, decurved, and brownish olive. Breeding adults mostly chestnut with deep purple on wings. Flies with quick wingbeats and intermittent gliding. Immature is duller than breeding adults, with head and neck streaked with white.
SIMILAR SPECIES. (1) Immature White Ibis has white belly, rump and upper-tail coverts; red bill; yellowish brown legs. (2) Limpkin is spotted with white on upper body and streaked below. In flight, its wingbeats are slower.

RANGE. Eastern United States, Costa Rica, Venezuela, Africa, tropical regions of Eurasia and Australia; Greater Antilles (except Jamaica). **Status**: Uncommon, but local, permanent and winter resident; also a transient, restricted to a few swamps such as Zapata and Birama. It also occurs on Isla de Pinos at Ciénaga de Lanier, and Cayos Romano and Coco. **Habitat**: Swamps and flooded ground such as rice fields. **Nesting**: May–Aug. Nests on trees at moderate heights. Builds a substantial shallow structure of twigs and sticks, sometimes with leaves and green stems in the lining. Lays three or four rather dark-blue eggs. **Voice**: Generally silent; occasionally a repeated grunt, *ka-onk*, and a low *kruk*. **Food**: Aquatic invertebrates.

ROSEATE SPOONBILL
SEVIYA; ESPÁTULA; CUCHARETA
Ajaia ajaja   Pl. 6

DESCRIPTION. 32" (81 cm). A large pink and white wader, with long spatulate gray bill. Immature is totally white at first, later increasingly pink; bill yellowish. Feeds by sweeping the bill through soft sediment while wading. Flies with slow wingbeats, sometimes soars. Commonly seen in small flocks.

SIMILAR SPECIES. (1) Scarlet Ibis and (2) White Ibis are smaller, with thin decurved bills and black wing tips. (3) Greater Flamingo is much larger, pink to vermillion all over, with very long neck and legs and extensive black on wings.

RANGE. Southeastern United States to Chile and Argentina; Bahamas, Cuba, and Hispaniola. **Status**: Common permanent resident on Cuba, and several cays; less abundant on Isla de Pinos. **Habitat**: Saltwater or freshwater swamps, marshes, mudflats, lagoons. **Nesting**: Oct–Jun. Nest is bulky structure of sticks and twigs. Breeds in secluded colonies among mangroves. Lays three white eggs with reddish brown spots. **Voice**: Silent except at breeding sites, where low-toned grunts are heard. **Food**: Crustacean, insects, aquatic plants.

PLATES

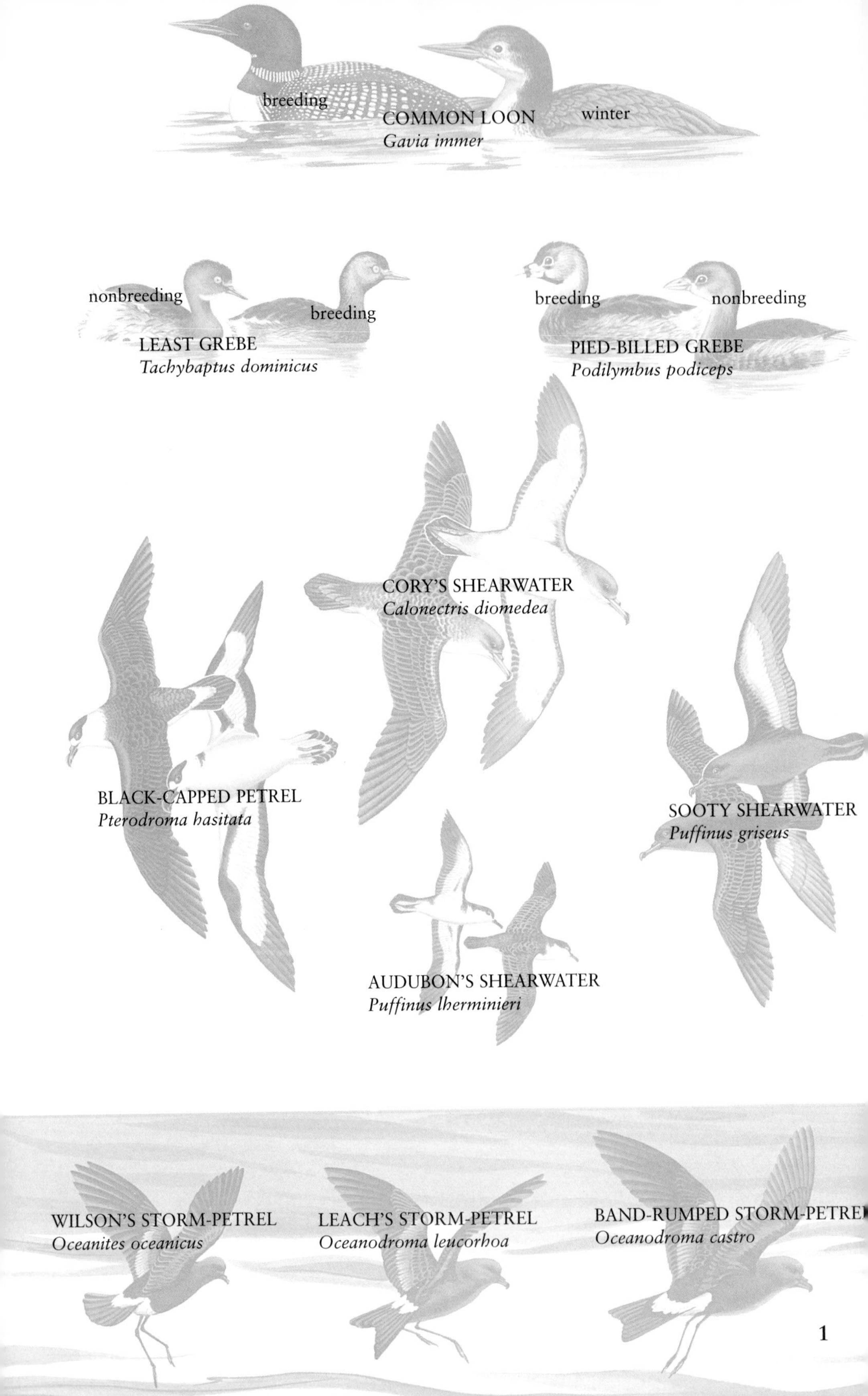

breeding

COMMON LOON
Gavia immer

winter

nonbreeding

breeding

LEAST GREBE
Tachybaptus dominicus

breeding

nonbreeding

PIED-BILLED GREBE
Podilymbus podiceps

CORY'S SHEARWATER
Calonectris diomedea

BLACK-CAPPED PETREL
Pterodroma hasitata

SOOTY SHEARWATER
Puffinus griseus

AUDUBON'S SHEARWATER
Puffinus lherminieri

WILSON'S STORM-PETREL
Oceanites oceanicus

LEACH'S STORM-PETREL
Oceanodroma leucorhoa

BAND-RUMPED STORM-PETREL
Oceanodroma castro

1

immature

MAGNIFICENT FRIGATEBIRD
Fregata magnificens

♀

♂

immature

WHITE-TAILED TROPICBIRD
Phaethon lepturus

adult

immature

RED-BILLED TROPICBIRD
Phaethon aethereus

adult

nonbreeding

AMERICAN WHITE PELICAN
Pelecanus erythrorhynchos

chick-feeding adult

juvenile

nonbreeding

BROWN PELICAN
Pelecanus occidentalis

AMERICAN WHITE PELICAN
Pelecanus erythrorhynchos

breeding

adult
breeding

MASKED BOOBY
Sula dactylatra

immature

adult

BROWN BOOBY
Sula leucogaster

immature

adult

nonbreeding

DOUBLE-CRESTED CORMORANT
Phalacrocorax auritus

RED-FOOTED BOOBY
Sula sula

NEOTROPIC CORMORANT
Phalacrocorax brasilianus

breeding

♂

ANHINGA
Anhinga anhinga

♀

breeding

ANHINGA
Anhinga anhinga

3

white morph

dark morph

immature

REDDISH EGRET
Egretta rufescens

GREAT BLUE HERON
Ardea herodias

dark morph

white morph

juvenile

immature

adult

TRICOLORED HERON
Egretta tricolor

immature

YELLOW-CROWNED NIGHT-HERON
Nyctanassa violacea

adult

adult

immature

BLACK-CROWNED NIGHT-HERON
Nycticorax nycticorax

adult

juvenile

AMERICAN BITTERN
Botaurus lentiginosus

♂

LEAST BITTERN
Ixobrychus exilis

♀

4

GREAT EGRET
Ardea alba

breeding

immature

adult

SNOWY EGRET
Egretta thula

breeding

adult

CATTLE EGRET
Bubulcus ibis

immature

immature

LITTLE BLUE HERON
Egretta caerulea

adult

adult

GREEN HERON
Butorides virescens

immature

5

SCARLET IBIS
Eudocimus ruber

WOOD STORK
Mycteria americana

immature

adult

breeding

GLOSSY IBIS
Plegadis falcinellus

nonbreeding

ROSEATE SPOONBILL
Ajaia ajaja

immature

adult

GREATER FLAMINGO
Phoenicopterus ruber

adult

WHITE IBIS
Eudocimus albus

adult

immature

dark "Blue Goose" morph

SNOW GOOSE
Chen caerulescens

white morph

CANADA GOOSE
Branta canadensis

adult

GREATER WHITE-FRONTED GOOSE
Anser albifrons

TUNDRA SWAN
Cygnus columbianus

adult

7

WEST INDIAN WHISTLING-DUCK
Dendrocygna arborea
adult

WHITE-FACED WHISTLING-DUCK
Dendrocygna viduata
adult

FULVOUS WHISTLING-DUCK
Dendrocygna bicolor
adult

BLACK-BELLIED WHISTLING-DUCK
Dendrocygna autumnalis
adult

GREEN-WINGED TEAL
Anas crecca
♂ ♀

GREEN-WINGED TEAL
Anas crecca
♀ ♂

MALLARD
Anas platyrhynchos
♂ ♀

MALLARD
Anas platyrhynchos
♀ ♂

WHITE-CHEEKED PINTAIL
Anas bahamensis

WHITE-CHEEKED PINTAIL
Anas bahamensis

NORTHERN PINTAIL
Anas acuta
♂ ♀

NORTHERN PINTAIL
Anas acuta
♀ ♂

8

BLUE-WINGED TEAL
Anas discors

♂ ♀

BLUE-WINGED TEAL
Anas discors

♀ ♂

CINNAMON TEAL
Anas cyanoptera

♂ ♀

CINNAMON TEAL
Anas cyanoptera

♀ ♂

NORTHERN SHOVELER
Anas clypeata

♂ ♀

NORTHERN SHOVELER
Anas clypeata

♀ ♂

GADWALL
Anas strepera

♂ ♀

GADWALL
Anas strepera

♀ ♂

AMERICAN WIGEON
Anas americana

♂ ♀

AMERICAN WIGEON
Anas americana

♀ ♂

WOOD DUCK
Aix sponsa

♂ ♀

WOOD DUCK
Aix sponsa

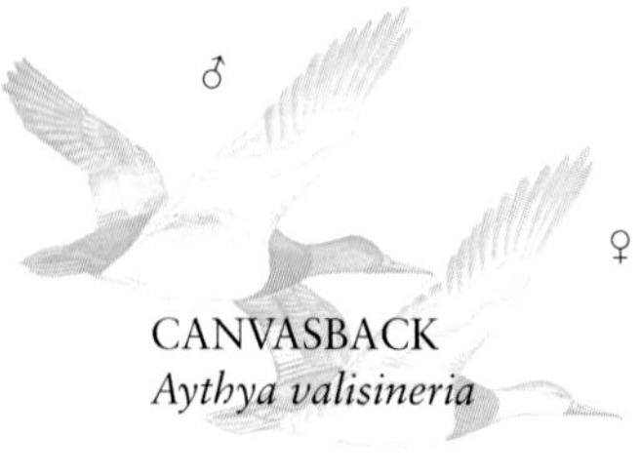

♀ ♂

CANVASBACK
Aythya valisineria

♂ ♀

CANVASBACK
Aythya valisineria

♀ ♂

REDHEAD
Aythya americana

♂ ♀

REDHEAD
Aythya americana

♀ ♂

9

RING-NECKED DUCK
Aythya collaris

RING-NECKED DUCK
Aythya collaris

LESSER SCAUP
Aythya affinis

LESSER SCAUP
Aythya affinis

BUFFLEHEAD
Bucephala albeola

BUFFLEHEAD
Bucephala albeola

HOODED MERGANSER
Lophodytes cucullatus

HOODED MERGANSER
Lophodytes cucullatus

RED-BREASTED MERGANSER
Mergus serrator

RED-BREASTED MERGANSER
Mergus serrator

RUDDY DUCK
Oxyura jamaicensis

immature

breeding

RUDDY DUCK
Oxyura jamaicensis

breeding

MASKED DUCK
Nomonyx dominicus

juvenile

breeding

MASKED DUCK
Nomonyx dominicus

breeding **10**

BLACK VULTURE
Coragyps atratus

TURKEY VULTURE
Cathartes aura

adult

immature

SNAIL KITE
Rostrhamus sociabilis

♂

♀

SWALLOW-TAILED KITE
Elanoides forficatus

adult

♀

♂

HOOK-BILLED KITE
Chondrohierax uncinatus

♀

11

adult

immature

SHARP-SHINNED HAWK
Accipiter striatus

adult

adult

GUNDLACH'S HAWK
Accipiter gundlachi

immature

immature

immature

BROAD-WINGED HAWK
Buteo platypterus

immature

adult

RED-TAILED HAWK
Buteo jamaicensis

immature

adult

COMMON BLACK-HAWK
Buteogallus anthracinus

adult

immature

spring

♀

NORTHERN HARRIER
Circus cyaneus

winter

♂

♂

♀

OSPREY
Pandion haliaetus

P. h. carolinensis

P. h. ridgwayi

♂

♂

P. h. carolinensis

12

immature

PEREGRINE FALCON
Falco peregrinus

adult
fall

CRESTED CARACARA
Caracara plancus

immature

adult

♀

MERLIN
Falco columbarius

♂

♂

AMERICAN KESTREL
Falco sparverius

immature

American race

white morph

♀

white morph

♀

red morph

♂

red morph

AMERICAN KESTREL
Falco sparverius

♀

NORTHERN BOBWHITE
Colinus virginianus

♂

HELMETED GUINEAFOWL
Numida meleagris

13

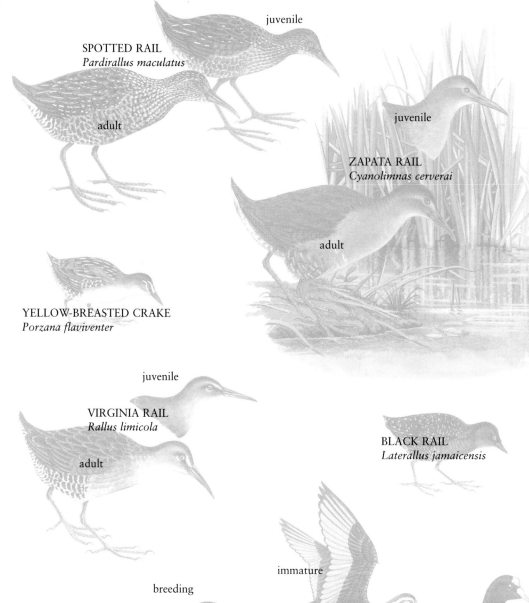

SPOTTED RAIL
Pardirallus maculatus

juvenile

adult

ZAPATA RAIL
Cyanolimnas cerverai

juvenile

adult

YELLOW-BREASTED CRAKE
Porzana flaviventer

juvenile

VIRGINIA RAIL
Rallus limicola

adult

BLACK RAIL
Laterallus jamaicensis

immature

breeding

SORA
Porzana carolina

winter

NORTHERN JACANA
Jacana spinosa

adult

14

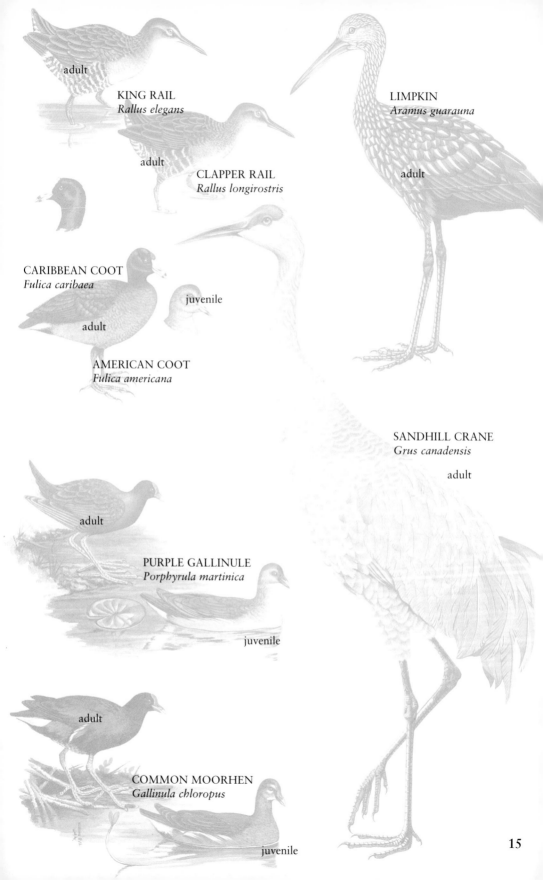

KING RAIL
Rallus elegans

adult

CLAPPER RAIL
Rallus longirostris

adult

LIMPKIN
Aramus guarauna

adult

CARIBBEAN COOT
Fulica caribaea

juvenile

adult

AMERICAN COOT
Fulica americana

SANDHILL CRANE
Grus canadensis

adult

adult

PURPLE GALLINULE
Porphyrula martinica

juvenile

adult

COMMON MOORHEN
Gallinula chloropus

juvenile

15

winter

SEMIPALMATED PLOVER
Charadrius semipalmatus

adult

♂

breeding

AMERICAN OYSTERCATCHER
Haematopus palliatus

SNOWY PLOVER
Charadrius alexandrinus

PIPING PLOVER
Charadrius melodus

winter

breeding

winter

♂

♀

AMERICAN AVOCET
Recurvirostra americana

WILSON'S PLOVER
Charadrius wilsonia

KILLDEER
Charadrius vociferus

AMERICAN GOLDEN-PLOVER
Pluvialis dominica

winter

adult

♂

BLACK-NECKED STILT
Himantopus mexicanus

BLACK-BELLIED PLOVER
Pluvialis squatarola

breeding

winter

16

WILSON'S PHALAROPE
Phalaropus tricolor
winter

RED-NECKED PHALAROPE
Phalaropus lobatus
winter

RED PHALAROPE
Phalaropus fulicaria
winter

WHIMBREL
Numenius phaeopus
adult

LONG-BILLED CURLEW
Numenius americanus
winter

HUDSONIAN GODWIT
Limosa haemastica
winter

MARBLED GODWIT
Limosa fedoa
winter

GREATER YELLOWLEGS
Tringa melanoleuca
winter

SOLITARY SANDPIPER
Tringa solitaria
winter

LESSER YELLOWLEGS
Tringa flavipes
winter

RUDDY TURNSTONE
Arenaria interpres
winter
breeding

UPLAND SANDPIPER
Bartramia longicauda
adult

17

winter

RED KNOT
Calidris canutus

winter

SANDERLING
Calidris alba

juvenile

winter

DUNLIN
Calidris alpina

winter

SEMIPALMATED SANDPIPER
Calidris pusilla

winter

PECTORAL SANDPIPER
Calidris melanotos

winter

WESTERN SANDPIPER
Calidris mauri

winter

LEAST SANDPIPER
Calidris minutilla

winter

STILT SANDPIPER
Calidris himantopus

winter

WHITE-RUMPED SANDPIPER
Calidris fuscicollis

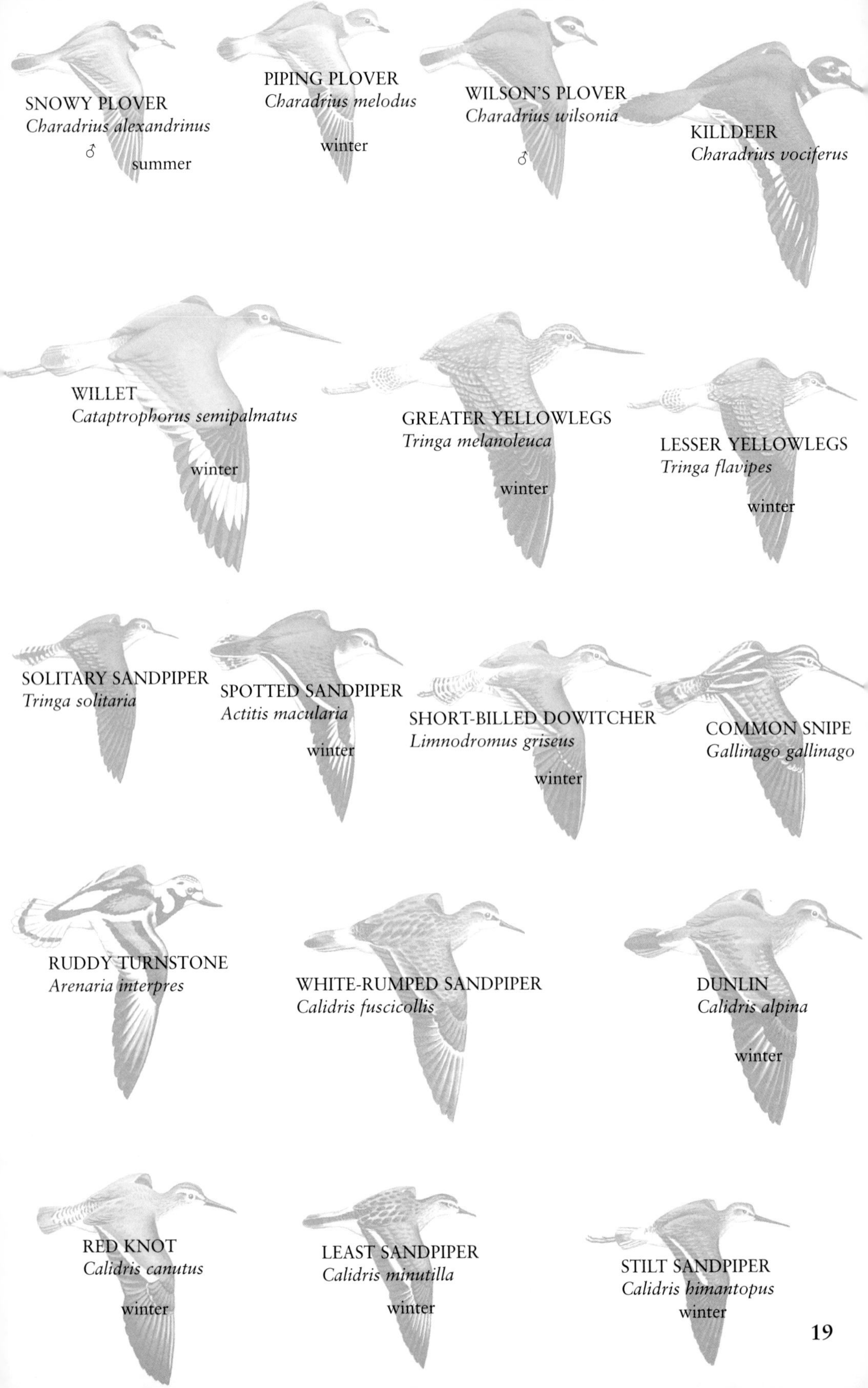

SNOWY PLOVER
Charadrius alexandrinus
♂
summer

PIPING PLOVER
Charadrius melodus
winter

WILSON'S PLOVER
Charadrius wilsonia
♂

KILLDEER
Charadrius vociferus

WILLET
Cataptrophorus semipalmatus
winter

GREATER YELLOWLEGS
Tringa melanoleuca
winter

LESSER YELLOWLEGS
Tringa flavipes
winter

SOLITARY SANDPIPER
Tringa solitaria

SPOTTED SANDPIPER
Actitis macularia
winter

SHORT-BILLED DOWITCHER
Limnodromus griseus
winter

COMMON SNIPE
Gallinago gallinago

RUDDY TURNSTONE
Arenaria interpres

WHITE-RUMPED SANDPIPER
Calidris fuscicollis

DUNLIN
Calidris alpina
winter

RED KNOT
Calidris canutus
winter

LEAST SANDPIPER
Calidris minutilla
winter

STILT SANDPIPER
Calidris himantopus
winter

19

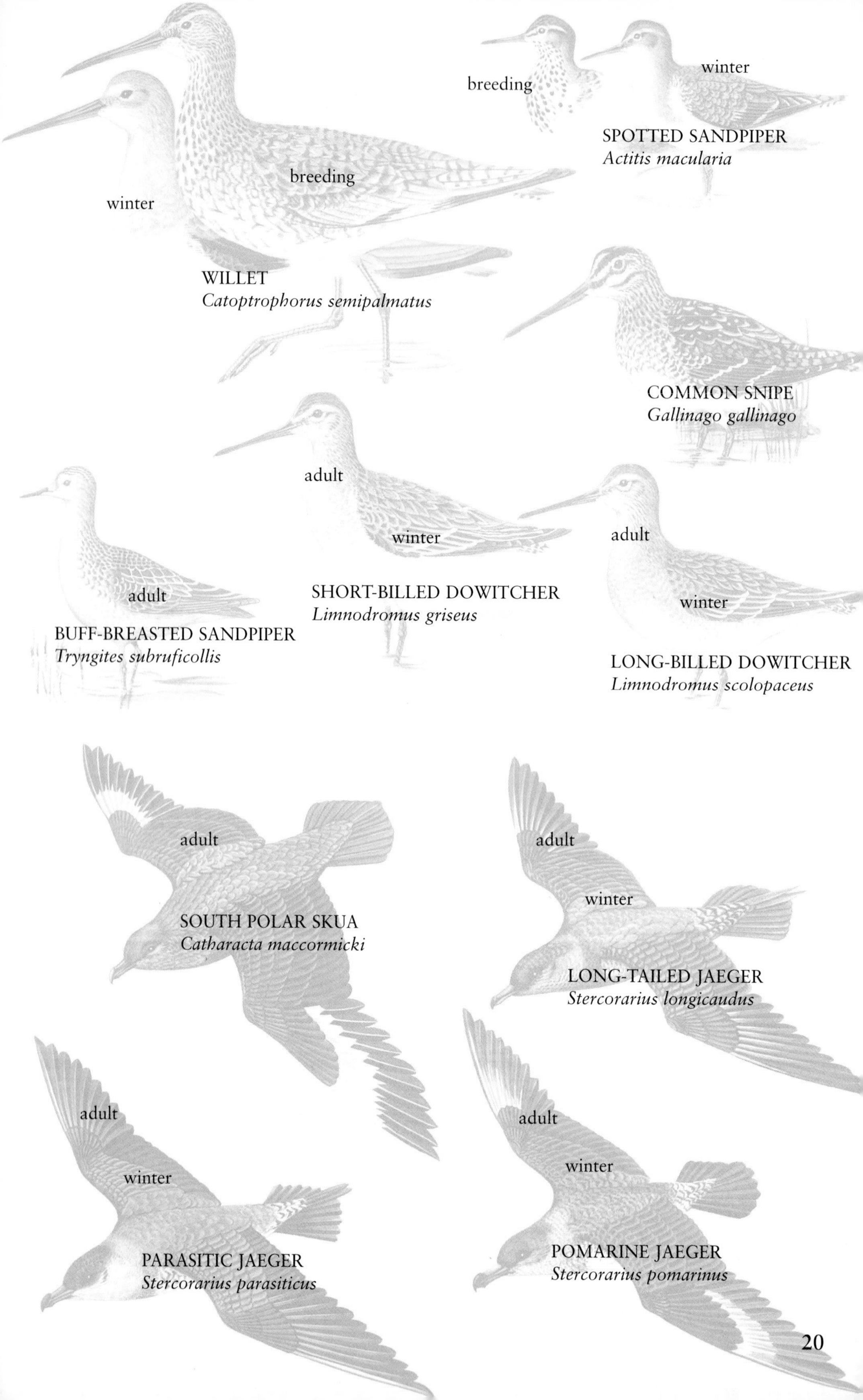

winter

breeding

SPOTTED SANDPIPER
Actitis macularia

breeding

winter

WILLET
Catoptrophorus semipalmatus

COMMON SNIPE
Gallinago gallinago

adult

winter

SHORT-BILLED DOWITCHER
Limnodromus griseus

adult

winter

LONG-BILLED DOWITCHER
Limnodromus scolopaceus

adult

adult

BUFF-BREASTED SANDPIPER
Tryngites subruficollis

adult

SOUTH POLAR SKUA
Catharacta maccormicki

adult

winter

LONG-TAILED JAEGER
Stercorarius longicaudus

adult

winter

PARASITIC JAEGER
Stercorarius parasiticus

adult

winter

POMARINE JAEGER
Stercorarius pomarinus

20

adult

winter

BLACK-HEADED GULL
Larus ridibundus

adult

winter

BLACK-LEGGED KITTIWAKE
Rissa tridactyla

1st winter

adult

winter

SABINE'S GULL
Xema sabini

adult

winter

BONAPARTE'S GULL
Larus philadelphia

breeding

LAUGHING GULL
Larus atricilla

2nd winter

LAUGHING GULL
Larus atricilla

juvenile

breeding

RING-BILLED GULL
Larus delawarensis

1st winter

RING-BILLED GULL
Larus delawarensis

breeding

1st winter

HERRING GULL
Larus argentatus

HERRING GULL
Larus argentatus

3rd winter

3rd winter

1st winter

GREAT BLACK-BACKED GULL
Larus marinus

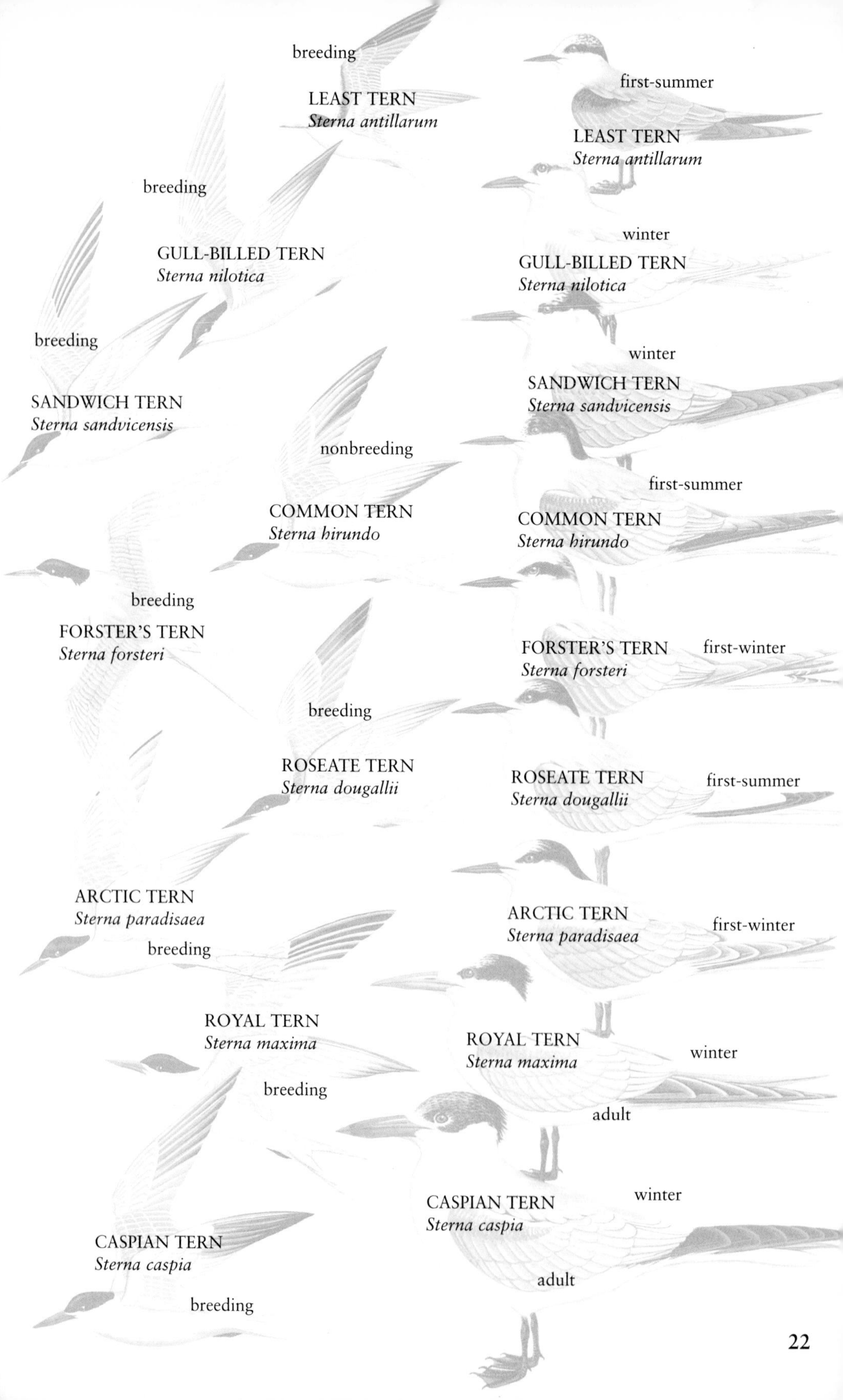

breeding

LEAST TERN
Sterna antillarum

first-summer

LEAST TERN
Sterna antillarum

breeding

GULL-BILLED TERN
Sterna nilotica

winter

GULL-BILLED TERN
Sterna nilotica

breeding

SANDWICH TERN
Sterna sandvicensis

winter

SANDWICH TERN
Sterna sandvicensis

nonbreeding

COMMON TERN
Sterna hirundo

first-summer

COMMON TERN
Sterna hirundo

breeding

FORSTER'S TERN
Sterna forsteri

FORSTER'S TERN
Sterna forsteri

first-winter

breeding

ROSEATE TERN
Sterna dougallii

ROSEATE TERN
Sterna dougallii

first-summer

ARCTIC TERN
Sterna paradisaea

ARCTIC TERN
Sterna paradisaea

first-winter

breeding

ROYAL TERN
Sterna maxima

ROYAL TERN
Sterna maxima

winter

breeding

adult

CASPIAN TERN
Sterna caspia

winter

CASPIAN TERN
Sterna caspia

adult

breeding

22

winter
BLACK TERN
Chlidonias niger

breeding
BLACK TERN
Chlidonias niger

adult
BRIDLED TERN
Sterna anaethetus

juvenile
BRIDLED TERN
Sterna anaethetus

adult
SOOTY TERN
Sterna fuscata

juvenile
SOOTY TERN
Sterna fuscata

adult
BROWN NODDY
Anous stolidus

juvenile
BROWN NODDY
Anous stolidus

adult
LARGE-BILLED TERN
Phaetusa simplex

juvenile
LARGE-BILLED TERN
Phaetusa simplex

juvenile
BLACK SKIMMER
Rhynchops niger

winter
DOVEKIE
Alle alle

adult

23

adult
♂

adult
♀

WHITE-CROWNED PIGEON
Columba leucocephala

juvenile

adult
♂

adult

MOURNING DOVE
Zenaida macroura

adult
♀

juvenile

adult

SCALY-NAPED PIGEON
Columba squamosa

adult

PLAIN PIGEON
Columba inornata

ROCK DOVE
Columba livia

adult
♀

adult

ZENAIDA DOVE
Zenaida aurita

♂

adult

KEY WEST QUAIL-DOVE
Geotrygon chrysia

♀ adult

♂ adult

WHITE-WINGED DOVE
Zenaida asiatica

♀

♂

COMMON GROUND-DOVE
Columbina passerina

♀ adult

adult ♂

GRAY-HEADED QUAIL-DOVE
Geotrygon caniceps

adult ♀

juvenile

adult ♂

RUDDY QUAIL-DOVE
Geotrygon montana

♀

♂

adult

adult

BLUE-HEADED QUAIL-DOVE
Starnoenas cyanocephala

adult

25

CUBAN MACAW
Ara tricolor

♂

PASSENGER PIGEON
Ectopistes migratorius

♀

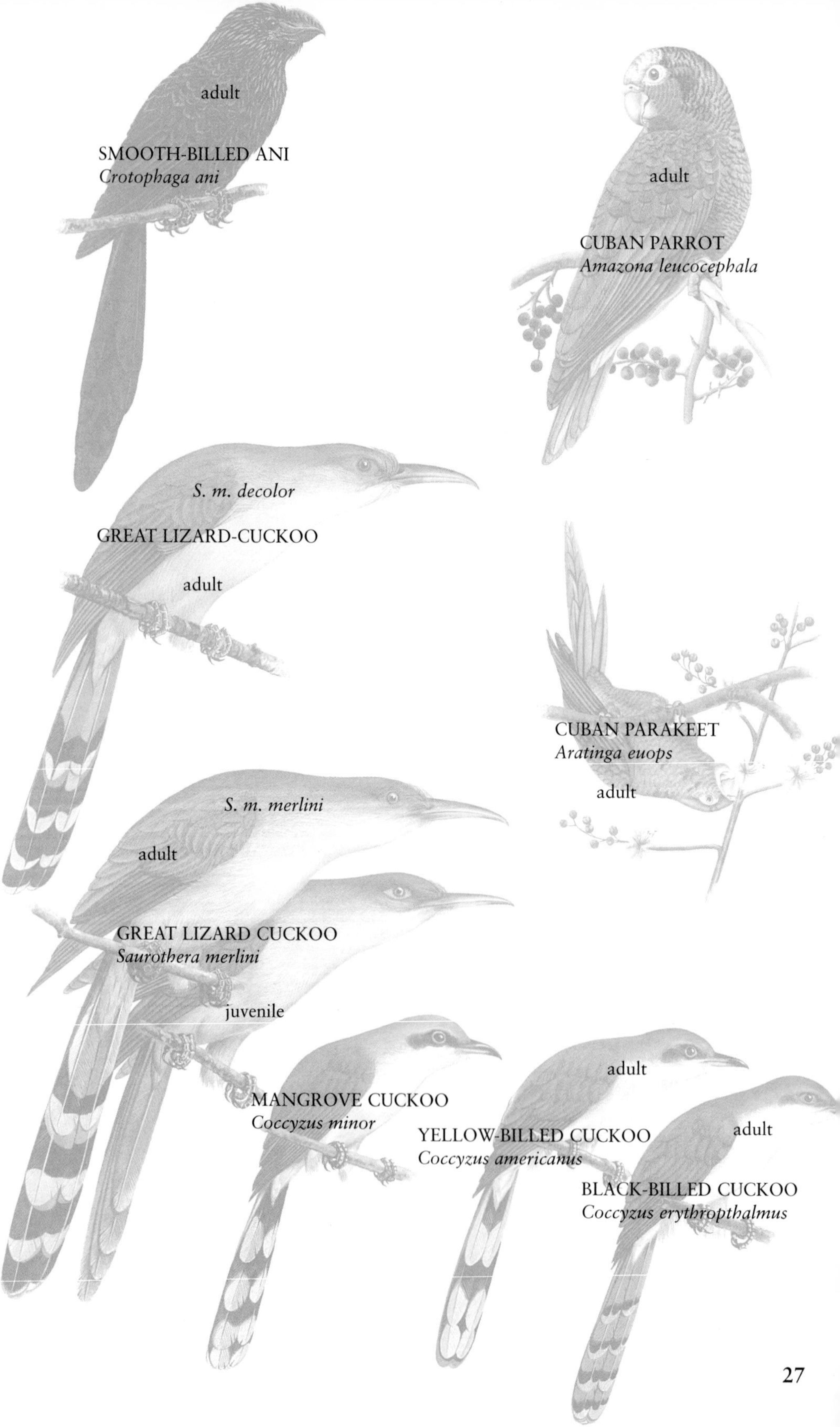

SMOOTH-BILLED ANI
Crotophaga ani

adult

CUBAN PARROT
Amazona leucocephala

adult

S. m. decolor

GREAT LIZARD-CUCKOO

adult

CUBAN PARAKEET
Aratinga euops

adult

S. m. merlini

adult

GREAT LIZARD CUCKOO
Saurothera merlini

juvenile

MANGROVE CUCKOO
Coccyzus minor

adult

YELLOW-BILLED CUCKOO
Coccyzus americanus

adult

BLACK-BILLED CUCKOO
Coccyzus erythropthalmus

27

BARN OWL
Tyto alba

SHORT-EARED OWL
Asio flammeus

LONG-EARED OWL
Asio otus

adult

STYGIAN OWL
Asio stygius

adult

CUBAN SCREECH-OWL
Otus lawrencii

CUBAN PYGMY-OWL
Glaucidium siju

adult

BURROWING OWL
Athene cunicularia

28

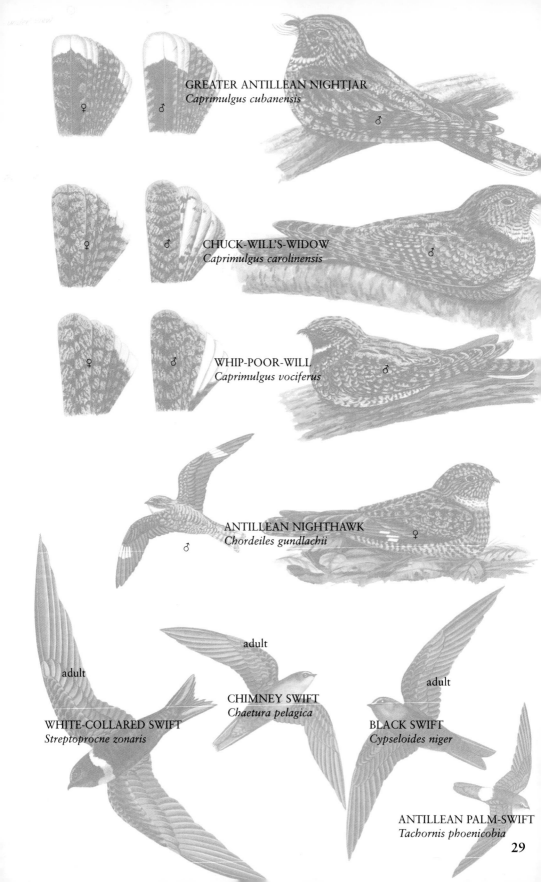

GREATER ANTILLEAN NIGHTJAR
Caprimulgus cubanensis

♀ ♂ ♂

CHUCK-WILL'S-WIDOW
Caprimulgus carolinensis

♀ ♂ ♂

WHIP-POOR-WILL
Caprimulgus vociferus

♀ ♂ ♂

ANTILLEAN NIGHTHAWK
Chordeiles gundlachii

♂ ♀

adult

WHITE-COLLARED SWIFT
Streptoprocne zonaris

adult

CHIMNEY SWIFT
Chaetura pelagica

adult

BLACK SWIFT
Cypseloides niger

adult

ANTILLEAN PALM-SWIFT
Tachornis phoenicobia

29

RUBY-THROATED HUMMINGBIRD
Archilochus colubris

CUBAN TROGON
Priotelus temnurus

immature

BEE HUMMINGBIRD
Mellisuga helenae

CUBAN EMERALD
Chlorostilbon ricordii

juvenile

juvenile

BELTED KINGFISHER
Ceryle alcyon

CUBAN TODY
Todus multicolor

juvenile

adult

30

juvenile

YELLOW-BELLIED SAPSUCKER
Sphyrapicus varius

♀ adult

♂ adult

adult

NORTHERN FLICKER
Colaptes auratus

♀

♂

adult ♂

♀ adult

CUBAN GREEN WOODPECKER
Xiphidiopicus percussus

juvenile

♂

WEST INDIAN WOODPECKER
Melanerpes superciliaris

♀

♀

♂

IVORY-BILLED WOODPECKER
Campephilus principalis

♀

♂

FERNANDINA'S FLICKER
Colaptes fernandinae

adult

FORK-TAILED FLYCATCHER
Tyrannus savana

adult ♂

SCISSOR-TAILED FLYCATCHER
Tyrannus forficatus

GREAT CRESTED FLYCATCHER
Myiarchus crinitus

adult

EASTERN WOOD-PEWEE
Contopus virens

LA SAGRA'S FLYCATCHER
Myiarchus sagrae

adult

CUBAN PEWEE
Contopus caribaeus

adult

WESTERN WOOD-PEWEE
Contopus sordidulus

adult

YELLOW-BELLIED FLYCATCHER
Empidonax flaviventris

adult

EASTERN PHOEBE
Sayornis phoebe

adult

ACADIAN FLYCATCHER
Empidonax virescens

adult

WILLOW FLYCATCHER
Empidonax traillii

adult

ALDER FLYCATCHER
Empidonax alnorum

32

TROPICAL KINGBIRD
Tyrannus melancholicus

EASTERN KINGBIRD
Tyrannus tyrannus

adult

GIANT KINGBIRD
Tyrannus cubensis

WESTERN KINGBIRD
Tyrannus verticalis

adult

LOGGERHEAD KINGBIRD
Tyrannus caudifasciatus

adult

GRAY KINGBIRD
Tyrannus dominicensis

juvenile

BLUE-HEADED VIREO
Vireo solitarius

adult

RED-EYED VIREO
Vireo olivaceus

adult

YELLOW-THROATED VIREO
Vireo flavifrons

adult

BLACK-WHISKERED VIREO
Vireo altiloquus

adult

WHITE-EYED VIREO
Vireo griseus

PHILADELPHIA VIREO
Vireo philadelphicus

adult

THICK-BILLED VIREO
Vireo crassirostris

WARBLING VIREO
Vireo gilvus

CUBAN VIREO
Vireo gundlachii

color variation
from Isla de Pinos

34

CUBAN CROW
Corvus nasicus

PALM CROW
Corvus palmarum

ZAPATA WREN
Ferminia cerverai

HOUSE WREN
Troglodytes aedon

TREE SWALLOW
Tachycineta bicolor

adult

winter

juvenile

CAVE SWALLOW
Hirundo fulva

adult

BAHAMA SWALLOW
Tachycineta cyaneoviridis

NORTHERN ROUGH-WINGED SWALLOW
Stelgidopteryx serripennis

♀

CUBAN MARTIN
Progne cryptoleuca

♂

adult

CLIFF SWALLOW
Hirundo pyrrhonota

♀

PURPLE MARTIN
Progne subis

adult

BARN SWALLOW
Hirundo rustica

juvenile

BANK SWALLOW
Riparia riparia

SWAINSON'S THRUSH
Catharus ustulatus

GRAY-CHEEKED THRUSH
Catharus minimus

WOOD THRUSH
Hylocichla mustelina

VEERY
Catharus fuscescens

♀

♂

EASTERN BLUEBIRD
Sialia sialis

adult

winter

NORTHERN WHEATEAR
Oenanthe oenanthe

juvenile

CUBAN GNATCATCHER
Polioptila lembeyei

adult

CUBAN SOLITAIRE
Myadestes elisabeth

adult

♀

late-winte

♂

RUBY-CROWNED KINGLET
Regulus calendula

BLUE-GRAY GNATCATCHER
Polioptila caerulea

adult

winter

CEDAR WAXWING
Bombycilla cedrorum

EUROPEAN STARLING
Sturnus vulgaris

37

AMERICAN ROBIN
Turdus migratorius

♂

GRAY CATBIRD
Dumetella carolinensis

T. p. rubripes

T. p. plumbeus

RED-LEGGED THRUSH
Turdus plumbeus

adult

adult

juvenile

adult

BAHAMA MOCKINGBIRD
Mimus gundlachi

NORTHERN MOCKINGBIRD
Mimus polyglottos

adult

BROWN THRASHER
Toxostoma rufum

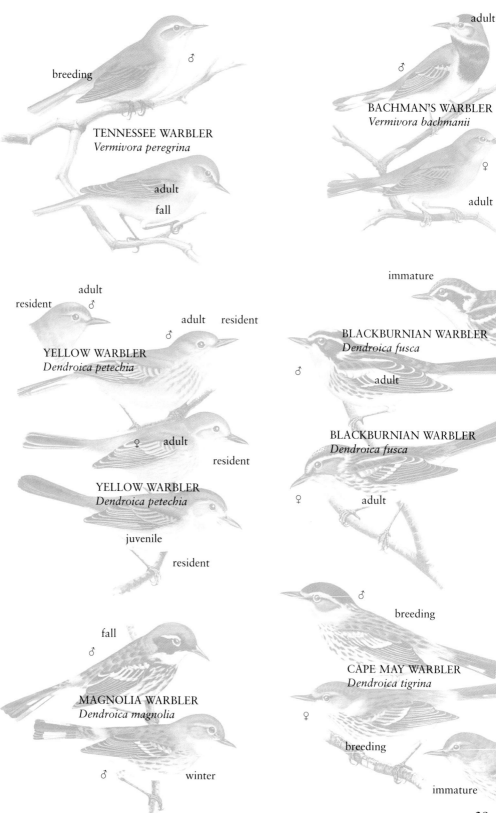

breeding

♂

TENNESSEE WARBLER
Vermivora peregrina

adult

fall

adult

♂

BACHMAN'S WARBLER
Vermivora bachmanii

♀

adult

resident

adult

♂

YELLOW WARBLER
Dendroica petechia

adult   resident

♂

adult

♀

YELLOW WARBLER
Dendroica petechia

resident

juvenile

resident

immature

BLACKBURNIAN WARBLER
Dendroica fusca

♂

adult

BLACKBURNIAN WARBLER
Dendroica fusca

♀   adult

fall

♂

MAGNOLIA WARBLER
Dendroica magnolia

♂   winter

♂   breeding

CAPE MAY WARBLER
Dendroica tigrina

♀

breeding

immature

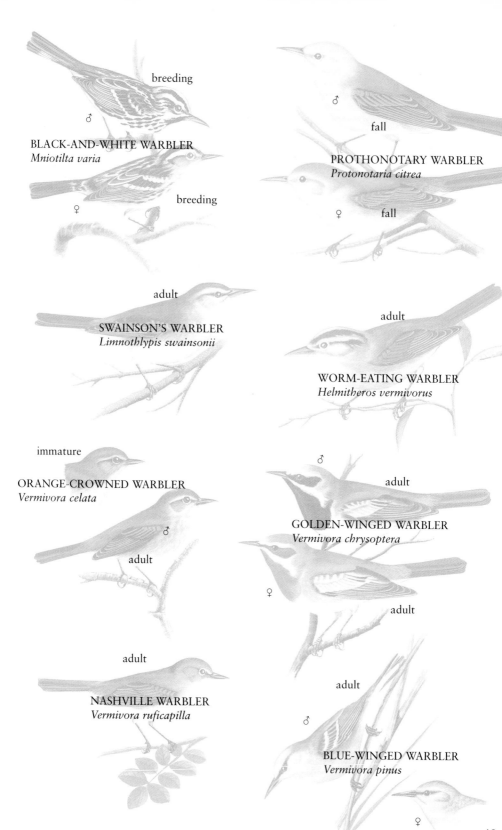

breeding

♂

BLACK-AND-WHITE WARBLER
Mniotilta varia

breeding

♀

♂

fall

PROTHONOTARY WARBLER
Protonotaria citrea

♀   fall

adult

SWAINSON'S WARBLER
Limnothlypis swainsonii

adult

WORM-EATING WARBLER
Helmitheros vermivorus

immature

ORANGE-CROWNED WARBLER
Vermivora celata

♂

adult

♂

adult

GOLDEN-WINGED WARBLER
Vermivora chrysoptera

♀

adult

adult

NASHVILLE WARBLER
Vermivora ruficapilla

adult

♂

BLUE-WINGED WARBLER
Vermivora pinus

♀

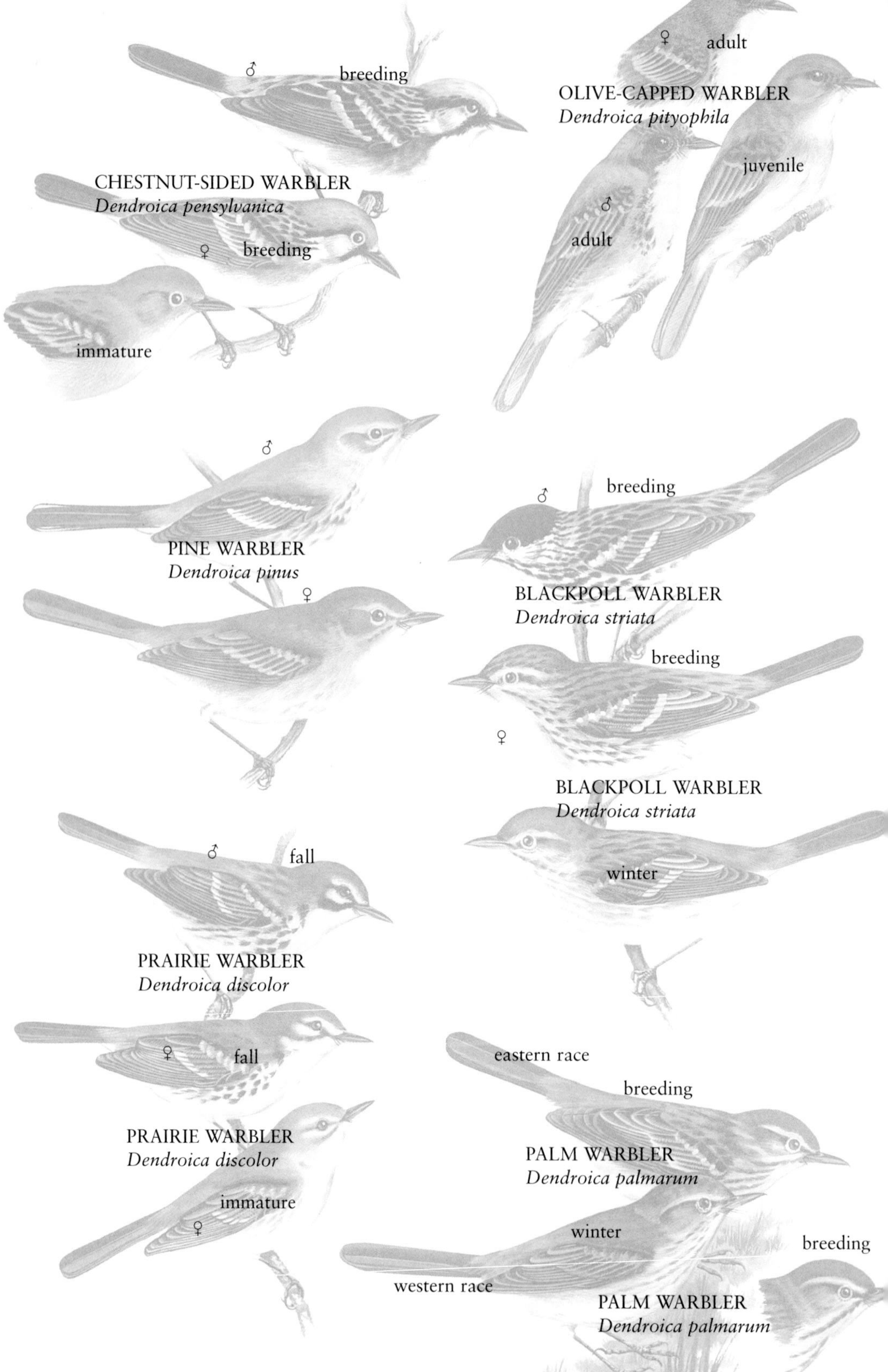

CHESTNUT-SIDED WARBLER
Dendroica pensylvanica

♂ breeding

♀ breeding

immature

OLIVE-CAPPED WARBLER
Dendroica pityophila

♀ adult

juvenile

♂ adult

PINE WARBLER
Dendroica pinus

♂

♀

BLACKPOLL WARBLER
Dendroica striata

♂ breeding

breeding

♀

BLACKPOLL WARBLER
Dendroica striata

winter

PRAIRIE WARBLER
Dendroica discolor

♂ fall

♀ fall

PRAIRIE WARBLER
Dendroica discolor

immature
♀

PALM WARBLER
Dendroica palmarum

eastern race
breeding

winter

western race

PALM WARBLER
Dendroica palmarum

breeding

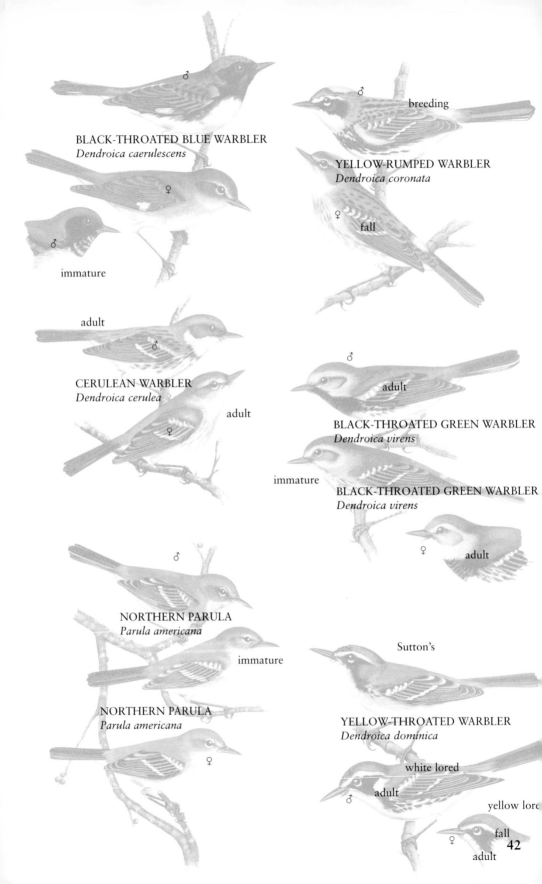

BLACK-THROATED BLUE WARBLER
Dendroica caerulescens

♂

♀

♂ immature

breeding

YELLOW-RUMPED WARBLER
Dendroica coronata

♀ fall

adult
♂

CERULEAN WARBLER
Dendroica cerulea

adult
♀

♂

adult

BLACK-THROATED GREEN WARBLER
Dendroica virens

immature

BLACK-THROATED GREEN WARBLER
Dendroica virens

♀ adult

♂

NORTHERN PARULA
Parula americana

immature

NORTHERN PARULA
Parula americana

♀

Sutton's

YELLOW-THROATED WARBLER
Dendroica dominica

white lored

♂ adult

yellow lore

♀ fall

42

adult

BAY-BREASTED WARBLER
Dendroica castanea

♂ breeding

adult

winter

♀ immature

COMMON YELLOWTHROAT
Geothlypis trichas

♂ immature

♂ adult

COMMON YELLOWTHROAT
Geothlypis trichas

YELLOW-BREASTED CHAT
Icteria virens

♀ adult

♀ adult

MOURNING WARBLER
Oporornis philadelphia

♀

KENTUCKY WARBLER
Oporornis formosus

♂

♂ adult

MOURNING WARBLER   immature
Oporornis philadelphia

S. n. noveboracensis

LOUISIANA WATERTHRUSH
Seiurus motacilla

S. n. notabilis

NORTHERN WATERTHRUSH
Seiurus noveboracensis

OVENBIRD
Seiurus aurocapillus

YELLOW-HEADED WARBLER
Teretistris fernandinae

43

adult

AMERICAN REDSTART
Setophaga ruticilla

♂

adult ♀

ORIENTE WARBLER
Teretistris fornsi

♂

HOODED WARBLER
Wilsonia citrina

♀

wing lining

breeding

♂

RED-LEGGED HONEYCREEPER
Cyanerpes cyaneus

♀

♀

WILSON'S WARBLER
Wilsonia pusilla

♂

adult

BANANAQUIT
Coereba flaveola

immature

♂

♀

fall

CANADA WARBLER
Wilsonia canadensis

44

STRIPE-HEADED TANAGER
Spindalis zena

♀

adult

♂

adult

SUMMER TANAGER
Piranga rubra

♀

♀

♂

SCARLET TANAGER
Piranga olivacea

breeding

early spring

♂

BOBOLINK
Dolichonyx oryzivorus

adult

fall

45

♂

♀
PAINTED BUNTING
Passerina ciris

immature

PAINTED BUNTING
Passerina ciris

INDIGO BUNTING
Passerina cyanea

♂

♀

breeding

♂

LAZULI BUNTING
Passerina amoena

♀

adult

♂

adult

CUBAN GRASSQUIT
Tiaris canora

♂

YELLOW-FACED GRASSQUIT
Tiaris olivacea

♀

adult

adult

adult

♂

♀

BLACK-FACED GRASSQUIT
Tiaris bicolor

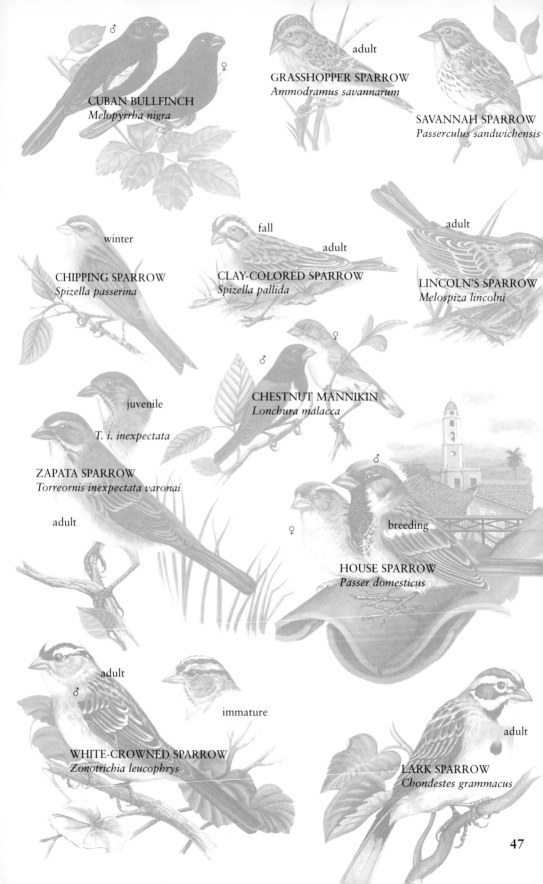

CUBAN BULLFINCH
Melopyrrha nigra

♂ ♀

adult

GRASSHOPPER SPARROW
Ammodramus savannarum

SAVANNAH SPARROW
Passerculus sandwichensis

winter

CHIPPING SPARROW
Spizella passerina

fall

adult

CLAY-COLORED SPARROW
Spizella pallida

adult

LINCOLN'S SPARROW
Melospiza lincolni

juvenile

T. i. inexpectata

ZAPATA SPARROW
Torreornis inexpectata varonai

adult

♂

♀

CHESTNUT MANNIKIN
Lonchura malacca

♂

♀

breeding

HOUSE SPARROW
Passer domesticus

adult

♂

immature

WHITE-CROWNED SPARROW
Zonotrichia leucophrys

adult

LARK SPARROW
Chondestes grammacus

EASTERN MEADOWLARK
Sturnella magna

SHINY COWBIRD
Molothrus bonariensis

♂

♀

♀

BROWN-HEADED COWBIRD
Molothrus ater

♂

breeding

♂

DICKCISSEL
Spiza americana

♀

ROSE-BREASTED GROSBEAK
Pheucticus ludovicianus

♂

♂

ROSE-BREASTED GROSBEAK
Pheucticus ludovicianus

♀

adult

GREEN-TAILED TOWHEE
Pipilo chlorurus

♀

♂

BLUE GROSBEAK
Guiraca caerulea

adult

♀

adult

RED-SHOULDERED BLACKBIRD
Agelaius assimilis

adult ♂

immature ♂

YELLOW-HEADED BLACKBIRD
Xanthocephalus xanthocephalus

♂

♀

♂

adult

GREATER ANTILLEAN GRACKLE
Quiscalus niger

♀

adult

♂

♀

TAWNY-SHOULDERED BLACKBIRD
Agelaius humeralis

adult

adult

CUBAN BLACKBIRD
Dives atroviolacea

49

first spring

♂

ORCHARD ORIOLE
Icterus spurius

♂

♀

♂

BLACK-COWLED ORIOLE
Icterus dominicensis

immature

adult

♀

HOODED ORIOLE
Icterus cucullatus

adult

♂

first spring

HOODED ORIOLE
Icterus cucullatus

♂

♀

BALTIMORE ORIOLE
Icterus galbula

adult

NORTHERN GANNET
Morus bassanus

MUSCOVY DUCK
Cairina moschata

♀

GREATER SCAUP
Aythya marila

♂

RING-NECKED PHEASANT
Phasianus colchicus

♀

EURASIAN COLLARED-DOVE
Streptopelia decaocto

MARSH WREN
Cistothorus palustris

adult

♂ winter

WESTERN TANAGER
Piranga ludoviciana

♀ winter

fall

RUSTY BLACKBIRD
Euphagus carolinus

♀ winter

AMERICAN GOLDFINCH
Carduelis tristis

♂ winter

51

STORKS Ciconiidae

Large wading birds with thick, long, usually decurved bills, stout at base. Neck and legs long, wings long and wide. Head partially or completely unfeathered. All fly with neck extended and slow wingbeats, frequently gliding. Storks live in swamps or wet grasslands, and feed on a large variety of vertebrates and crustaceans. Sexes alike. (**W**:26; **C**:1)

WOOD STORK
CAYAMA
Mycteria americana Pl. 6

DESCRIPTION. 41″ (104 cm). Very large, white, with dark naked head. Black flight feathers and tail. Immature has paler head, yellow bill, and grayish plumage. Often soars high.

SIMILAR SPECIES. (1) White Ibis is much smaller, has black only on wing tips, and red bill and legs. (2) Sandhill Crane is larger with all-gray wings and red crown.

RANGE. Southern United States to northern Argentina; Cuba, Dominican Republic. **Status**: Rare permanent resident on Cuba, Isla de Pinos, and the larger of the northern cays. **Habitat**: Mangroves, brackish lagoons. **Nesting**: Jan–Apr. A colonial breeder. Nest like that of herons, on mangrove or other tall trees. Lays three or four creamy-white eggs. **Voice**: Usually silent. Occasionally a harsh croak. **Food**: Vertebrates, crustaceans.

AMERICAN VULTURES Cathartidae

Large black birds with small naked heads, long broad wings, and hooked bills. Often soar high, alone or in large groups, circling within thermals. Food is mainly carrion, located by keen sight and smell. Sexes alike. (**W**:7; **C**:2)

BLACK VULTURE
ZOPILOTE

Coragyps atratus   Pl. 11

DESCRIPTION. 25" (64 cm). Entirely black, with gray head and legs. Tail short and square. In flight, white patches show on wing tips; when soaring, wings are outstretched horizontally. Flaps rapidly, with short glides.

SIMILAR SPECIES. Immature Turkey Vulture has flight feathers silvery below, and more soaring flight; when soaring, wings held in a shallow V shape. Tail longer and rounded.

RANGE. Southern and eastern United States to central Argentina. **Status**: Very rare. Ten sight records: 1928; 1940; 1960; 1961; 1962; 1967; 1970; 1979; 1995; 11 Jul (1996). Western and central part of the island. **Habitat**: Coastal and open areas. **Voice**: Usually silent. May emit some grunts. **Food**: Carrion.

TURKEY VULTURE
AURA TIÑOSA

Cathartes aura   Pl. 11

DESCRIPTION. 29" (74 cm). Very dark chocolate brown, appearing black from a distance, with red head. Soars with a characteristic rocking motion, with wings slightly above the horizontal, forming a shallow V; silvery flight feathers evident. Often seen in groups of 50 or more. In early morning or after rain, often perches with outstretched wings. Immature has grayish head.

SIMILAR SPECIES. (1) Black Vulture has white patches on outer wing, flatter flight profile, and shorter and more square-shaped tail. (2) Male Snail Kite has finely hooked bill, feathered head, and a broad white band on tail base.

RANGE. Southern Canada to southern Argentina; Greater Antilles, introduced (?) in Puerto Rico. **Status**: Abundant permanent resident in Cuba, Isla de Pinos, and the larger cays. Also a rare winter resident. **Habitat**: Open country, towns, swamps, and forest. **Nesting**: Jan–Jun. Builds no nest. Lays one to three grayish-white, spotted eggs, on rocky cliffs, in

caves, sugar cane fields, or on forest floor.**Voice**: Usually silent. May produce some grunts when disturbed. **Food**: Mainly carrion, also garbage at
dumps.

FLAMINGOS Phoenicopteridae

Among the most spectacular of birds, these large, highly social waders have
very long legs and neck. The peculiar bill is thick and abruptly decurved.
Flight fast and steady. They feed on very small animals and plants strained
from sediments while bill is under water. Sexes alike. (**W**:5; **C**:1)

GREATER FLAMINGO
FLAMENCO
Phoenicopterus ruber   Pl. 6

DESCRIPTION. 45" (114 cm). The tallest of Cuban birds. Entirely pink to
  vermillion, except for black flight feathers, usually visible only in flight;
  pink bill with black tip. Immature is white to pink, with paler bill. Flies
  with neck and legs fully extended.
SIMILAR SPECIES. Roseate Spoonbill is smaller, much paler, with spatulate
  bill and shorter neck; lacks black on wings.
RANGE. Southern Bahamas, Cuba, Hispaniola, Gonâve and Beata islands,
  Netherland Antilles, Yucatán peninsula, Galápagos, locally in northern
  South America. Widespread in the Old World. **Status**: Common permanent resident along Cuban coast, mainly in the Archipiélago de Sabana-
  Camagüey, slowly declining in numbers. Easily seen at La Salina in the Zapata peninsula. Not reported from the Archipiélago las Coloradas, and
  only a few in Archipiélago de los Canarreos. **Habitat**: Inshore shallows,
  brackish ponds. **Nesting**: Apr–Oct. Breeds in large colonies of many hundreds or thousands, in shallow, very muddy, isolated coastal wetlands.
  Nests are built on wet ground, of mud mixed with marine grasses and
  twigs. The single immaculate white egg is placed in a depression atop the

30-cm-high, roughly conical nest mound. **Voice**: Loud gooselike honking. **Food**: Small invertebrates, algae.

SWANS, GEESE, AND DUCKS Anatidae

Medium-sized water birds, with broad flattened bills and webbed feet. Legs usually short. Wings are pointed; tail usually short. Most species found in Cuba are winter visitors, some accidental; six are resident. Excellent swimmers, they feed in the water and on land, mainly on aquatic plants. Sexes often differ. Many are social. (**W**:157; **C**:29)

FULVOUS WHISTLING-DUCK
YAGUASÍN; YAGUASA MEXICANA
Dendrocygna bicolor   Pl. 8

DESCRIPTION. 20" (51 cm). Cinnamon brown; wings and back darker, almost black; white stripes along sides. White rump is conspicuous in flight. Bill and legs gray. As in other species of whistling-ducks, legs are long, with feet extending beyond the tail in flight. Takes off vertically from water; in flight, head and neck are lowered. Feeds while walking, often in rice fields. Immature resembles adult, but duller.

SIMILAR SPECIES. West Indian Whistling-Duck is larger, with longer neck and more rounded head; cheek pale; sides mottled with white; no white on rump.

RANGE. Southern United States, Mexico, Honduras, Costa Rica; South America; eastern Africa, Madagascar, India, Sri Lanka and Burma; Cuba, Hispaniola, and Puerto Rico. **Status**: Common permanent resident and migrant on Cuba and Isla de Pinos. **Habitat**: Rice fields, recently plowed or flooded. Also small lagoons. **Nesting**: Apr–Dec. Builds nest with rice plants, grasses, and reeds. More than one female may lay eggs in a single nest. Single-female clutch is 11 to 18 pale-yellow eggs. **Voice**: When taking flight, a high-pitched double whistle. **Food**: Vegetable matter, seeds, grasses, rice.

WEST INDIAN WHISTLING-DUCK
YAGUASA; CUBA LIBRE
Dendrocygna arborea Pl. 8

DESCRIPTION. 21″ (53 cm). Upperparts, neck, and head dark brown, with dark stripe from hindcrown to back. Belly much paler. Whitish chin and throat. Sides mottled with white. Bill, legs, rump, and tail blackish. Immature similar to adult, but duller.

SIMILAR SPECIES. (1) Fulvous Whistling-Duck is smaller, cinnamon brown with white stripes on sides and white band on rump. (2) Female Northern Pintail is more uniformly brown, has blackish speculum with iridescent purple and green sheen, and has white belly and long, pointed tail.

RANGE. Greater Antilles (including Cuba, Isla de Pinos, Grand Cayman, and Île-à-Vache off Hispaniola), the Bahamas (Andros, San Salvador, and Inagua islands), and northern Lesser Antilles (at least on Barbuda and Antigua). **Status**: Common permanent resident in Cuba, Isla de Pinos, and certain cays. Vulnerable. **Habitat**: Swamps, mangroves, and lagoons. Occasionally roosts and even breeds in coastal areas. **Nesting**: Apr–Dec. Builds crude nests in trees, laying clutches of 11 to 16 white eggs. **Voice**: In flight, a pleasant whistle, said to sound like the phrase *Cuba libre*; when feeding in groups, emits a soft murmur. **Food**: Seeds, fruit, tender shoots. Royal palm fruits are relished.

WHITE-FACED WHISTLING-DUCK
YAGUASA CARIBLANCA
Dendrocygna viduata Pl. 8

DESCRIPTION. 17″ (43 cm). Brown back, black wings, reddish chestnut chest. Back of head and neck black; crown and cheeks white; white spot on patch. Black bill; gray legs. Immature has entirely gray head and neck, and pale gray belly.

SIMILAR SPECIES. Black-bellied Whistling-Duck has an all-gray face, black belly, and a large white wing patch.

RANGE. Costa Rica through most of South America, Africa, Madagascar, and the Comoro Islands. **Status**: Vagrant. Four records: Apr, Jun (1858);

1937; 23 Apr (1947). Ponds in Zapata peninsula and San Cristobal. **Habitat**: Freshwater lagoons. **Voice**: More or less prolonged whistles: *swee-ree-ree*. **Food**: Seeds, grain, insects.

BLACK-BELLIED WHISTLING-DUCK
YAGUASA BARRIGUIPRIETA; YAGUASA PECHINEGRA
Dendrocygna autumnalis   Pl. 8

DESCRIPTION. 20" (51 cm). Crown, hindneck, chest, and back cinnamon brown. Sides of head and throat gray; belly, rump, tail, and trailing edge of wings black. Wings have broad white stripe. Legs and bill pink, orange, or red. A striking bird in flight, its black belly and broad white wing stripe being very distinct. Immature is duller with gray bill.

SIMILAR SPECIES. White-faced Whistling-Duck has a white mask, black bill, and gray legs.

RANGE. Southwestern United States and Mexico to northern Argentina. **Status**: Apparently a permanent resident in Cuba. Very rare. Recently established in rice fields of Ciego de Ávila province. **Habitat**: Rice fields, swamps, lagoons. **Nesting**: Undescribed in Cuba. **Voice**: A shrilling *pe-che-wee-che* whistle. **Food**: Insects, grain, seeds.

TUNDRA SWAN
CISNE
Cygnus columbianus   Pl. 7

DESCRIPTION. 52" (132 cm). White, with long neck and black bill and legs. Most birds have a yellow spot in front of eye. Immature is pale ashy gray with pink bill, dusky at tip.

SIMILAR SPECIES. Snow Goose is smaller, with shorter neck and pink bill.

RANGE. Arctic coasts of North America and Eurasia, wintering in New World south to Atlantic and Pacific coasts of United States. **Status**: Vagrant. Two records: 17 Dec (1944); 25 Feb (1990). Zapata peninsula, Nipe. **Habitat**: Freshwater ponds. **Voice**: Loud, high-pitched yelping. **Food**: Aquatic vegetation, grass, insects.

GREATER WHITE-FRONTED GOOSE
GUANANA; GANSO

Anser albifrons Pl. 7

DESCRIPTION. 29" (73 cm). Grayish brown. Underparts with irregular dark barring. Pink bill, with base surrounded by white feathers. Lower belly and tail coverts white. Legs yellow to orange. First-winter immature similar to adult but without barring on underparts.

SIMILAR SPECIES. Immature dark-morph Snow Goose has dark bill and legs.

RANGE. Circumpolar. North American populations winter to southern United States, western North America, and Mexico. **Status**: Vagrant. One record: Zapata peninsula, 1946. Gundlach considered it a rather common winter resident in the 19th century. **Habitat**: Freshwater ponds. **Voice**: A fast, loud, repeated laughter. **Food**: Aquatic insects and larvae.

SNOW GOOSE
GUANANA PRIETA; GANSO

Chen caerulescens Pl. 7

DESCRIPTION. 28" (71 cm). Two color morphs, white and dark ("Blue Goose"). In both, adults have pink bill and legs. Lower belly and undertail coverts white. White morph has black primaries and rust-colored feathers on face. Dark morph is grayish brown with white head and neck, black primaries and secondaries. Intermediates occur. Immature white morph has gray back with dark legs and bill; dark morph is almost entirely grayish brown, including bill and legs.

SIMILAR SPECIES. (1) Immature White-fronted Goose has pink or orange bill. (2) Tundra Swan is larger, with longer neck and black legs.

RANGE. Breeds in the Arctic; winters to southern United States and Mexico. **Status**: Vagrant. Four records: 13 Apr (1925); Jan (1985); 25 Apr; another undated. **Habitat**: Borders of swamps, flooded marshes, lagoons, agricultural land. **Voice**: A loud, resonant *whouk*. **Food**: Shoots and roots of sedges, aquatic invertebrates.

CANADA GOOSE
GANSO DE CANADÁ

Branta canadensis Pl. 7

DESCRIPTION. 25–45" (64–114 cm). Ashy brown. Head, long neck, legs, and bill black. A conspicuous white patch on cheek; belly and undertail coverts white. White band on rump.

SIMILAR SPECIES. None.

RANGE. Northern North America, wintering south to Mexico and northern Florida. **Status**: Vagrant. Two records: 11 Dec (1966); 17 Apr (1972). La Habana and Pinar del Río provinces. **Habitat**: Freshwater ponds. **Voice**: A deep and resonant *honk*. **Food**: Shoots, roots, grain.

MUSCOVY DUCK

PATO DOMÉSTICO

Cairina moschata   Pl. 51

DESCRIPTION. Male: 32" (80 cm); female: 25" (63 cm). A clumsy black
gooselike duck with large white wing patches and underwing coverts. *Male*:
has a bare, knobby red face. *Female*: duller; may lack lack facial knobs. Immature black with small white spot on upperwing. Domestic varieties can
be white, black, or mottled. Escaped birds often become semiwild.

SIMILAR SPECIES. None.

RANGE. Northern Mexico south through Central America to Argentina.
Status: Widely introduced. **Habitat**: Ponds, slow-moving rivers, often
near towns or farms. **Nesting**: May–Jul. Lays eight to nine greenish white
eggs in tree hollows. **Voice**: Usually silent, but may emit hissing noises or
low quacks. **Food**: Seeds, aquatic vegetation, fruits, insects, small fish.

WOOD DUCK

HUYUYO; PATO HUYUYO

Aix sponsa   Pl. 9

DESCRIPTION. 19" (48 cm). *Male*: head large and conspicuously crested,
dark iridescent green, violet, white, and black; bill blackish with extensive
red base. *Female*: brown, with purplish sheen on back, smaller crest; white
around eye. In both sexes, throat and belly are white. Eclipse male similar
to female, but with more white under chin; bill similar to breeding male.
Juvenile similar to female, with spotted belly; legs dull yellow. Takeoff
from water is vertical. Nods or pumps head when swimming. Often
perches in trees. In flight long, square tail and short neck are evident.

SIMILAR SPECIES. Female Hooded Merganser has slender bill, lacks white
around eye, has white patches on wings, and more direct flight.

RANGE. Southern Canada and United States, wintering south to Mexico,
Bahamas, and Cuba. **Status**: Uncommon permanent resident on Cuba
and Isla de Pinos. **Habitat**: Lagoons, lakes, estuaries; rice irrigation canals

and well-vegetated reservoirs. **Nesting**: Jul–Oct. In tree cavities, including dead palms. Lays bone-white eggs in clutches of 8 to 14. **Voice**: Female makes a high-pitched *whoo-eek* on takeoff. **Food**: Vegetable matter, seeds, aquatic insects.

GREEN-WINGED TEAL
PATO SERRANO; CERCETA

Anas crecca   Pl. 8

DESCRIPTION. 14" (36 cm). *Male*: gray, with chestnut head and neck, and a green patch from eye to nape. *Female*: mottled brown. Both sexes have green speculum bordered by buff leading edge, white belly, mostly white undertail coverts, and gray or olive feet. Bill very small. Immature resembles female. Flight very fast. In flocks this duck is prone to aerial acrobatics of twist, turn, and dive.
SIMILAR SPECIES. (1) Female Blue winged-Teal has a blue patch on forewing, larger bill, and yellow legs. (2) Female Cinnamon Teal also has a blue patch on forewing, and a more spatulate bill.
RANGE. Canada; northern, western, and central United States. Eurasia. North American populations winter south to Belize, Honduras, and West Indies. **Status**: Uncommon winter resident and transient in Cuba and Isla de Pinos (Aug–Apr). **Habitat**: Marshes, lagoons, rice fields, bays. **Voice**: Male: an abrupt, short whistle; female: a weak croak. **Food**: Grain, grass, seeds, plant shoots.

MALLARD
PATO INGLÉS

Anas platyrhynchos   Pl. 8

DESCRIPTION. 23" (58 cm). *Male*: back grayish brown, lighter on belly; chest and sides chestnut. Head dark green with bright metallic sheen and white neck ring. Bill yellow. *Female*: mottled yellowish brown, dark bill with base and orange tip. Both sexes have black tail with white outer tail feathers, violet blue speculum bordered with white, and orange legs. Juvenile is similar to female, with dull olive bill. Takeoff from water is vertical.
SIMILAR SPECIES. (1) Male Northern Shoveler is smaller, with long, wide black bill and white chest. Blue patch on forewing is conspicuous in flight. (2) Female Gadwall has white speculum bordered with black, and gray bill with orange sides.
RANGE. North America, Mexico, Europe, Central Asia, Japan. North American populations winter south to central Mexico. **Status**: Very rare transient and winter resident in Cuba and Isla de Pinos (Sep–Apr). **Habitat**: Freshwater ponds. **Voice**: A loud, resonant *quack*. **Food**: Seeds, grass, grain, insects.

WHITE-CHEEKED PINTAIL
PATO DE BAHAMAS; PATO GUINEO

Anas bahamensis   Pl. 8

DESCRIPTION. 17" (43 cm). *Male*: yellowish brown, spotted black. Throat, foreneck, and cheek white; base of bill scarlet. Speculum green, with buffy border. Tail long and pointed. *Female*: similar to male, but duller.

SIMILAR SPECIES. None.

RANGE. Bahamas, Greater Antilles, northern Lesser Antilles, South America. **Status**: Common, but local, permanent resident, mainly in rice fields south of Sancti-Spíritus, Birama. **Habitat**: Estuaries, brackish coastal lagoons, rice fields. **Nesting**: May–Aug. Nest built of grasses and aquatic vegetation on the ground near water. Lays 6 to 12 light-cream-colored eggs. **Voice**: Male: a squeak; female: a *quack*. **Food**: Grain, leaves, weeds.

NORTHERN PINTAIL
PATO PESCUECILARGO

Anas acuta   Pl. 8

DESCRIPTION. 26" (66 cm). Slim, with long neck and long pointed tail. Bill blue gray. Blackish speculum, with iridescent purple and green sheen, and a thin white border on its rear. Slender pointed wings. *Male*: gray,with dark brown head and white underparts; neck with a white lengthwise stripe. *Female*: mottled brown, with white belly conspicuous in flight. Immature resembles female. Takeoff from water is vertical.

SIMILAR SPECIES. (1) West Indian Whistling-Duck has darker wings and different flight profile, with head, neck, and legs carried lower than the rest of the body. Wingbeats slower. (2) Female Gadwall has white speculum bordered with black and gray bill with orange sides.

RANGE. North America, wintering south to West Indies and rarely to northern South America; Eurasia. **Status**: Common winter resident and transient in Cuba, Isla de Pinos, and some cays (Sep–Apr). **Habitat**: Shallow lagoons, brackish ponds, canals. **Voice**: Male: a rarely heard double whistle; female: a low croak. **Food**: Seeds, aquatic vegetation, insects, tadpoles.

BLUE-WINGED TEAL
PATO DE LA FLORIDA
Anas discors   Pl. 9

DESCRIPTION. 16" (41 cm). *Male*: ashy chestnut brown, densely spotted with black, with a conspicuous white crescent before eye; undertail coverts black. *Female*: mottled brown. Eclipse male and immature similar to female. In flight, both sexes show light blue patch on forewing. Male breeding plumage is acquired in November. This fast-flying duck twists and turns in small compact flocks.

SIMILAR SPECIES. (1) Female Cinnamon Teal has a slightly longer and more spatulate bill. Eye line less distinct. (2) Female Green-winged Teal lacks blue forewing patches. White belly distinct in flight. (3) Female Northern Shoveler has similar wing pattern in flight, but has much larger spatulate bill.

RANGE. North America, wintering south to central Argentina, and West Indies. **Status**: Abundant winter resident and transient in Cuba, Isla de Pinos, and some cays (25 Aug–12 Jun). **Habitat**: Lakes, lagoons, marshes, estuaries, brackish coastal lagoons, and coastal inlets with dense vegetation. **Voice**: Male: a whistling chirp; female: a weak *quack*. **Food**: Grain, aquatic invertebrates.

CINNAMON TEAL
PATO CANELO
Anas cyanoptera   Pl. 9

DESCRIPTION. 16" (41 cm). *Male*: in breeding plumage, cinnamon red with dark brown mottling on back. In flight, body appears very dark overall. *Female*: ashy brown, mottled and variegated. Both sexes have green iridescent speculum with black inner and outer borders and blue patch on forewing. Male in eclipse plumage is inseparable from females in the field. Takeoff from water is vertical.

SIMILAR SPECIES. (1) Female Blue-winged Teal has slightly shorter and less spatulate bill, more distinct loral spot and eye line; male has white crescent before eye and a white spot on flanks near tail. (2) Female Green-winged Teal lacks blue patch on forewing and has shorter bill. (3) Female Northern Shoveler has similar wing pattern in flight, but has much larger spatulate bill.

RANGE. Western North America, northern Mexico. Also resident in South America, wintering south through Central America to northern South America. **Status**: Vagrant. Three records: 9 Jan (1917); 5 Dec (1963); 3 Feb. Western part of the island. **Habitat**: Shallow freshwater with abundant vegetation. **Voice**: Male: a whistling *peep*; female: a harsh *karr*. **Food**: Aquatic vegetation, seeds, snails, insects.

NORTHERN SHOVELER
PATO CUCHARETA
Anas clypeata   Pl. 9

DESCRIPTION. 19" (48 cm). *Male*: white chest, chestnut underparts, dark-green iridescent head. *Female*: mottled brown. Both sexes have long, spoon-shaped bills, dark gray in male, and gray tinged with orange on cutting edges and lower mandible in female; a large pale blue patch on forewing, green speculum, and orange feet. Many immature males show a white crescent on the face. Swimming birds carry the bill angled downward toward the water. In flight, the heavy, downward-angled bill gives the bird a bulky appearance.

SIMILAR SPECIES. (1) Male Mallard has chestnut chest, white belly, and a shorter yellow bill. (2) Female Blue-winged and (3) Cinnamon Teals have similar wing pattern in flight but much smaller bills.

RANGE. Eurasia, North America. North American populations winter south to northern South America and West Indies. **Status**: Common winter resident and transient in Cuba, Isla de Pinos, and some larger cays (Sep–16 May). **Habitat**: Lagoons, ponds, flooded savannas with thick vegetation. **Voice**: Male: a deep guttural cackle; female: a weak *quack*. **Food**: Aquatic vegetation, snails, clams, aquatic insects.

GADWALL
PATO GRIS
Anas strepera   Pl. 9

DESCRIPTION. 20" (51 cm). *Male*: gray, with brown neck and head, black tail coverts. A dull chestnut patch on forewing is inconspicuous in flight. *Female*: mottled brown, with slightly paler neck and head. Both sexes have white speculum bordered with black, white belly, dull yellow legs, and gray upper mandible with orange sides. Takeoff from water is vertical. Often dives.

SIMILAR SPECIES. (1) Female American Wigeon has pale gray head, bluish bill, and green speculum, and in flight shows large white patch on forewing. (2) Mallard female has darker belly, less steep forehead, and violet blue speculum bordered with white. (3) Female Northern Pintail has longer neck, pointed tail, and gray bill.

RANGE. Eurasia and North America. North American populations winter in southern United States, Mexico, Bahamas, and Jamaica. **Status**: Vagrant. Three records: 13 Mar (1971); 7 Mar (1997); 8 Oct. Pinar del Río province, Zapata peninsula. **Habitat**: Lakes, lagoons. **Voice**: Male: an large repertoire of whistles, warblings, and growls; female: a low *quack*. **Food**: Aquatic plants, algae.

AMERICAN WIGEON

PATO LAVANCO

Anas americana Pl. 9

DESCRIPTION. 20" (51 cm). *Male*: brownish, with gray head; white crown and dark green patch on side of head. *Female*: brown, with entirely gray head and neck, and white belly. Both sexes have large white patches on forewing, more distinct in male than in female, and bluish gray, black-tipped bill. The speculum is green bordered with black. Often feeds on land. Takeoff from water is vertical.

SIMILAR SPECIES. Female Gadwall has small white speculum, rather than broad white patch on forewing; brownish head; and steeper forehead.

RANGE. North America, wintering in coastal and southern United States, south occasionally to Colombia and Venezuela; West Indies. **Status**: Common winter resident and transient in Cuba, Isla de Pinos, and some cays (Aug–18 May). **Habitat**: Lagoons, brackish coastal ponds. **Voice**: Male: a pleasant melodious whistle; female: a weak *quack*. **Food**: Grass, grain, aquatic vegetation.

CANVASBACK

PATO LOMIBLANCO

Aythya valisineria Pl. 9

DESCRIPTION. 22" (56 cm). *Male*: whitish, with dark chestnut-brown head, and black chest and tail. *Female*: pale brownish-gray back and sides, with brown head and neck. Both sexes have long dark blackish bills and sloping foreheads. Rump and tail coverts black; tail black. In flight, wings lack contrasting pattern. Patters when taking off from water. Flight very fast.

SIMILAR SPECIES. Male Redhead has gray back, short, tricolored bill, rounder head, and gray trailing edge of wing.

RANGE. Western and central North America, wintering to southern Mexico, Gulf Coast, and Florida. **Status**: Vagrant. Four records: 28 Dec (1951); 25 Nov, Nov (1987); 10 Feb (1988). Zapata peninsula and Pinar del Río province. **Habitat**: Large lagoons with dense shoreline vegetation. **Voice**: Male produces a grunting sound, and other short notes: *cu* or *mu*. **Food**: Aquatic vegetation, aquatic invertebrates.

REDHEAD

PATO CABECIRROJO

Aythya americana Pl. 9

DESCRIPTION. 20" (51 cm). *Male*: gray, with rounded, reddish brown head, and black chest, rump, and tail coverts. *Female*: tawny brown all over with whitish belly. Both sexes have blue bill with a white ring and black tip. Pale gray speculum with narrow white borders. Immature resembles female. Gray wing stripe evident in flight.

SIMILAR SPECIES. Male Canvasback has whitish back, long blackish bill, and sloping forehead.

RANGE. Western and southeastern Canada and northwestern and northeastern United States, wintering south to Guatemala; Bahamas and Jamaica. **Status**: Vagrant. Two records: 11 Nov; 5 Mar. Western provinces. **Habitat**: Freshwater lakes and lagoons. **Voice**: Male: a vibrant note, reminiscent of a mewing cat. **Food**: Aquatic vegetation, aquatic invertebrates.

RING-NECKED DUCK
PATO CABEZÓN; PATO MORISCO
Aythya collaris   Pl. 10

DESCRIPTION. 17" (43 cm). *Male*: blackish purple head, with gray sides and a white mark before the wing. Back black; belly white. *Female*: dark brown, whitish near bill, with white eye ring. Both sexes have a dark bill, bluish in the male, with a narrow white ring across the tip; and a small rounded crest, barely evident in the female. Wide gray stripe on wing is conspicuous in flight. Patters when taking off from the water.

SIMILAR SPECIES. Male Lesser Scaup has gray back, and a bold white stripe only on the secondaries. Bill pale blue, with black tip.

RANGE. North America, wintering south to Panama and West Indies. **Status**: Common winter resident and transient in Cuba, Isla de Pinos and some larger cays (Sep–Apr). **Habitat**: Lagoons, estuaries, canals, flooded savannas. **Voice**: Rarely heard, a sort of purring. Male occasionally produces a weak whistle. **Food**: Aquatic insects, snails.

GREATER SCAUP*
PATO CABEZÓN RARO
Aythya marila   Pl. 51

DESCRIPTION. 18" (46 cm). *Male*: simply patterned, with black breast, brownish black rump and tail. Back gray, finely barred; belly white; head and neck with a dark greenish gloss. *Female*: mostly brown, with a bold white patch surrounding base of bill. May have pale ear patch. Belly white, mottled dusky. Both sexes have pale blue bill and bold white wing stripe extending to wing tip, visible in flight. Immature similar to female, with a reduced pale face patch. Rises from water by pattering along the surface.

SIMILAR SPECIES. Lesser Scaup appears flatheaded and male has glossy purple head and neck. Female lacks whitish ear patch. In flight, white wing stripe does not extend to wing tip.

RANGE. Breeds in northern North America and northern Eurasia. North American populations winter to both Pacific and Atlantic coasts of North America. **Status**: One observed at Los Canales, Zapata peninsula, in the 1960s. **Habitat**: Brackish lagoons. **Voice**: Silent in winter. **Food**: Mollusks, aquatic plants.

LESSER SCAUP

PATO MORISCO; PATO NEGRO; PATO CABEZÓN

Aythya affinis   Pl. 10

DESCRIPTION. 17" (43 cm). *Male*: similar to Greater Scaup; slightly smaller with purplish black head and neck, black chest, brownish black tail and rump. *Female*: dark brown, with distinctive white patch surrounding base of bill. In spring, female may have distinct white ear patch. Both sexes have white belly and a broad white wing stripe, blue bill with black tip, and flat crown. In flight, shows a white wing stripe restricted to secondaries. Rises from water by pattering along the surface.

SIMILAR SPECIES. (1) Male Greater Scaup has greenish black head. Both sexes have a distinctly rounded head. In flight, shows white stripe on secondaries and most of primaries. (2) Male Ring-necked Duck has black back; in flight, broad gray wing stripe is visible. Female has white ring on bill and white eye ring and postocular line.

RANGE. Northwestern North America, wintering south to northern South America; West Indies. **Status**: Rare winter resident and transient in Cuba, Isla de Pinos, and Cayo Coco (Oct–21 Apr). **Habitat**: Flooded savannas, lagoons, lakes, freshwater ponds. **Voice**: Usually silent. **Food**: Aquatic invertebrates, aquatic vegetation.

BUFFLEHEAD

PATO MOÑUDO

Bucephala albeola   Pl. 10

DESCRIPTION. 14" (36 cm). *Male*: black back, white underparts. Head black, with a green or purple gloss and a large white patch that develops from smaller patches on sides of head in first-winter birds. Tail gray. *Female*: dark brown, with white chest and belly, and a small white patch on sides of head. Both sexes have very short bluish bill and white speculum, which extends across the entire wing in males. Immature resembles adult female.

SIMILAR SPECIES. Male Hooded Merganser is larger, with brown sides and long thin bill.

RANGE. Canada and northwestern United States, wintering to southern United States, Mexico; Greater Antilles. **Status**: Vagrant. Two records: one last century; another 3 Nov (1967). Gibara. **Habitat**: Freshwater ponds and lakes. **Voice**: Infrequently heard. Male produces guttural sounds; female, a sort of *quack*, and *guk*. **Food**: Aquatic snails, aquatic vegetation.

HOODED MERGANSER

PATO DE CRESTA; PATO SERRUCHO

Lophodytes cucullatus   Pl. 10

DESCRIPTION. 18" (48 cm). *Male*: black head, neck, and back; white breast, brown sides. *Female*: brown, darker on back. Both sexes have thin, black bill, the lower mandible yellowish in females. When erect, crest is high and rounded, and decorated with a fan-shaped white patch in males; in females, crest is rusty orange and bushy. Crest is flattened in flight.

SIMILAR SPECIES. (1) Red-breasted Merganser is larger, with long red bill. Male has all-green head. (2) Male Bufflehead is all white below, with rounded head and shorter, thick, bluish bill. (3) Female Wood Duck has white around eye, thicker bill, and lacks white wing patches.

RANGE. North America, wintering south to southeastern United States and Mexico; Bahamas and Puerto Rico. **Status**: Vagrant. Three records: 16 Nov–8 Dec (1996); 16 Jan; Feb. Cienfuegos, Cayo Coco. **Habitat**: Coasts, freshwater ponds, lakes, lagoons. **Voice**: Croaking notes. **Food**: Fish, crustaceans, insects.

RED-BREASTED MERGANSER
PATO SERRUCHO

Mergus serrator   Pl. 10

DESCRIPTION. 23" (58 cm). *Male*: black head and back, head having green gloss; reddish brown chest, streaked with black; white neck and large wing patch. *Female*: grayish, with brown head and hindneck; whitish throat and breast. Both sexes have long red bills and shaggy crest. Females more common in Cuba than males.

SIMILAR SPECIES. Male Hooded Merganser is smaller with prominent white crest flaring behind head. Female Hooded Merganser is almost entirely brown with rounded buffy crest.

RANGE. Northern North America and Eurasia. American populations winter along both coasts of North America south to Mexico. **Status**: Rare winter resident in Cuba and Cayo Coco (18 Dec–30 Apr). **Habitat**: Bays and open ocean. **Voice**: Usually silent outside the breeding season. **Food**: Fish, aquatic insects, crustaceans.

MASKED DUCK
PATO AGOSTERO; PATO CHICO; PATO CRIOLLO

Nomonyx dominicus   Pl. 10

DESCRIPTION. 13" (33 cm). *Male*: in breeding plumage, chestnut with black face and black-speckled sides and upperparts, and blue bill. *Female*: in breeding plumage, reddish brown with two black stripes across face; a black cap and dark bill. Both sexes have white wing patches. Nonbreeding male and female are similar to breeding female, but less warmly toned. Juvenile similar to nonbreeding female, but duller. Chicks also show distinct stripes on face and four round whitish spots on back. Pattering takeoff from water, although not commonly seen in flight. Typically, skulks in emergent vegetation and floats low in water; difficult to see, except at dusk and at night when it ventures into more open water.

SIMILAR SPECIES. Female and nonbreeding male Ruddy Duck have all-dark wing and a single stripe across face.

RANGE. Gulf coast of Texas to Central and South America; West Indies. **Status**: Widespread but rare permanent resident in Cuba and Isla de Pinos. Vulnerable. Most easily seen at Los Canales and Santo Tomás in the Zapata peninsula, where occasionally locally common. **Habitat**: Swamps, flooded savannas, lagoons, rice plantations, overgrown canals. **Nesting**: Jun–Sep. Nests at lagoon shores and rice fields, laying 8 to 18 cream-colored eggs. **Voice**: A short *du-du-du*, like a clucking hen. **Food**: Vegetable matter, aquatic insects.

RUDDY DUCK
PATO CHORIZO; PATO ROJO; PATO ESPINOSO
Oxyura jamaicensis   Pl. 10

DESCRIPTION. 15" (38 cm). *Male*: in breeding plumage, reddish brown with black cap, white cheek and blue bill. In nonbreeding plumage, similar to female but with white cheek, browner below, and heavily barred. *Female*: grayish brown with dark cap and pale cheek patch crossed by a dark line. Dark bill. In winter, grayer with a less distinct dark stripe on cheek. Patters when taking flight. Immature resembles female.

SIMILAR SPECIES. Female and nonbreeding male Masked Ducks have two stripes on face and white wing patch, and breeding male has black mask. Skulks in emergent vegetation; difficult to see.

RANGE. Canada, south to Guatemala; South America; West Indies. **Status**: Rare permanent resident, winter resident and transient in Cuba and Isla de Pinos. **Habitat**: Flooded savannas, rice fields, lagoons. **Nesting**: Jun–Aug. Nests in thick aquatic vegetation. Lays up to six whitish eggs. **Voice**: Male produces an emphatic clucking while courting. **Food**: Mainly vegetable matter.

KITES AND HAWKS Accipitridae

Medium-sized to large birds with strong hooked bills and powerful grip-ping feet. Females are larger than males and sometimes differently colored or patterned. Diurnal, they feed on live animals, from snails or insects to mam-mals. Some hunt from perches; others, in flight. Excellent fliers. Nest is usu-ally a rather crude platform high in trees. Sadly, larger species have a bad rep-utation as a menace to poultry and, if possible, are shot. Uncommon near towns. (**W**:240; **C**:10)

OSPREY
GUINCHO; GUARAGUAO DE MAR
Pandion haliaetus   Pl. 12

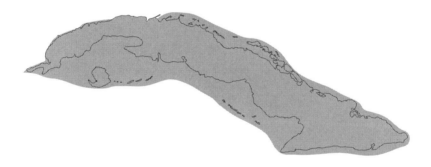

DESCRIPTION. 23" (58 cm). Dark brown above; white below. In flight the long wings have a conspicuous black patch on wrist, and black tips. Migra-tory individuals show broad dark stripe through the eye, almost black in male; residents are paler on back, with less distinct stripe on head. Female chest is lightly spotted with brown. Immature has dorsal and upperwing feathers with buffy edges; crown more heavily streaked. Takes fish by hov-ering and then diving feet-first into water. Has distinctive appearance in flight, with wings bent backward at wrist. Rarely soars.
SIMILAR SPECIES. None.
RANGE. All continents except Antarctica. South America only during the winter. West Indies. **Status**: Common permanent and winter resident; also

transient in Cuba, Isla de Pinos, and cays. **Habitat**: Shallow waters around cays, lagoons, and some inland man-made basins; less frequent along Cuba's coasts. **Nesting**: Oct–May. Builds huge nests of sticks on rather low tree stumps, or on rocky outcrops. Lays three cream eggs with chestnut red spots. The North American race is known to breed in Cuba (two records). **Voice**: A series of short, high-pitched whistles. **Food**: Fish captured in either freshwater or salt water.

HOOK-BILLED KITE
GAVILÁN CAGUARERO; GAVILÁN SONSO; GUARAGUAO

Chondrohierax uncinatus Pl. 11

DESCRIPTION. 15" (38 cm). *Male*: ash-colored, appearing whitish from a distance, barred below, mostly with gray. *Female*: dark brown, with conspicuous reddish bars on whitish chest and belly. Both sexes have long, banded tail, very heavy hooked bill. The cere and facial skin are blue. Eye of adult is bluish white. In flight, wings have broadly rounded tips and very distinct black and white barring in primaries. Immature female has head and upper back speckled with white. Wing feathers edged with rufous.

SIMILAR SPECIES. None.

RANGE. Southern Texas and Mexico south to northern Argentina and southern Peru. In the West Indies, only Grenada and Cuba. **Status**: A critically endangered endemic subspecies, that may warrant species status (ssp. *wilsonii*). Only three sightings in the past 30 years; most recent one at Yateras in 1992. **Habitat**: Reported by Gundlach in 1850 at Zapata peninsula and in 1876 at Cienfuegos, but has subsequently been found only in the humid mountains of the Grupo Sagua-Baracoa, Santiago de Cuba, and Guantánamo provinces. **Nesting**: Unknown. **Voice**: Unknown. **Food**: Apparently tree snails.

SWALLOW-TAILED KITE
GAVILÁN COLA DE TIJERA

Elanoides forficatus Pl. 11

DESCRIPTION. 23" (58 cm). Wings long and pointed, black from above, black and white from below. Back and tail black; head, chest, underwing

coverts, and belly white. Tail long and very deeply forked. Immature similar to adult, but duller, with some fine streaks on head and breast.
SIMILAR SPECIES. None.
RANGE. Southeastern United States to northern Argentina, wintering in South America. Occurs regularly in western Greater Antilles. **Status**: Uncommon transient, and possibly a rare winter resident in Cuba (Aug–Dec; Feb–Apr). **Habitat**: Swamps, savannas, river mouths, clearings. **Voice**: A loud shrill whistle: *ke-wee-wee* or *je-wee-wee*, the first note very short. **Food**: Reptiles (especially canopy-dwelling snakes), frogs, insects; also nestling birds. Usually eats in flight.

SNAIL KITE
GAVILÁN CARACOLERO; GAVILÁN BABOSERO
Rostrhamus sociabilis   Pl. 11

DESCRIPTION. 17ʺ (43 cm). *Male*: very dark brown, appearing almost black from a distance. Base of bill, eye, and legs orange red. *Female*: dark brown with densely streaked breast and belly. Forehead, throat, and eye-stripe whitish. Legs yellow to orange. Both sexes have long, finely hooked bill, a whitish patch on base of primaries, and broad white tail base. Immature resembles female, but with finer streaking on underparts and paler legs and eyes; back and wings with a scaly appearance caused by pale feather edges; washed buff below. Flight is slow and leisurely, interrupted by intermittent glides. Wings held horizontally, typically low with upturned primary tips. Tail appears triangular and is often twisted as bird positions itself in the air.
SIMILAR SPECIES. (1) Northern Harrier has narrower wings and white rump (not tail base). Flight profile is a shallow V. (2) Common Black-Hawk has banded tail, thick legs, heavier and straighter bill. (3) Turkey Vulture is entirely black with unfeathered pink or red head; usually soars in a shallow V.
RANGE. Southern Florida; Mexico to Argentina. **Status**: Common permanent resident in Cuba, occurring much less frequently on Isla de Pinos and some cays, such as Santa María. Commonly observed at Laguna del Tesoro and Laguna Grande. **Habitat**: Swamps, lagoons, flooded fields. **Nesting**: Apr–Jul. Nest low in a bush or tree. Builds a nest of twigs, varying from a

small flimsy platform to a compact structure. Lays three white eggs, spotted with reddish brown. **Voice**: A dry, rasping chatter, *ga-ga-ga-ga*. **Food**: Apple snails (*Pomacea*) are seized with feet from just under the surface of the water, and are taken to preferred feeding posts to be eaten.

NORTHERN HARRIER

GAVILÁN SABANERO; GAVILÁN DE CIÉNAGA

Circus cyaneus   Pl. 12

DESCRIPTION. 21" (53 cm). *Male*: pale gray to dark brown; very light below; tail faintly barred. *Female*: dark brown, with chest and belly whitish, streaked; tail with distinct banding. Immatures dark brown with plain cinnamon chest, belly, and wing coverts on leading edge of wing. Both sexes and immatures have long tail, white rump patch, long legs, and facial disks, giving the face an owllike appearance. Flies low and erratically and glides with slightly uptilted wings. Perches on ground rather than on posts. Roosts communally on the ground. Prey devoured at kill site.

SIMILAR SPECIES. (1) Gundlach's Hawk flies more directly and with a flatter flight profile. It also has unmarked rump and white undertail coverts. (2) Snail Kite has sharply hooked bill, broader rounded wings, and broad white tail base, holds wings horizontally.

RANGE. North America and Eurasia. North American populations winter south to West Indies and (rarely) northern South America. **Status**: Common winter resident and transient in Cuba, Isla de Pinos, and northern cays (15 Aug–24 Apr). **Habitat**: Flat, open terrain, such as savannas and marshes. **Voice**: Usually silent. **Food**: Rodents, reptiles, frogs, large insects, occasionally small birds.

SHARP-SHINNED HAWK

GAVILANCITO; HALCONCITO; FALCONCITO

Accipiter striatus   Pl. 12

DESCRIPTION. 12" (30 cm). Dark bluish gray above; whitish below with breast and belly barred with reddish brown. Light bodied, with long square tail and short, rounded wings. Head small, not extending far in front of wings during flight. Immature dark brown above, with streaked breast and belly; may have whitish spots on upperparts. Like other *Accipiter* species, flies with alternating bursts of shallow, rapid wingbeats and glides.

SIMILAR SPECIES. Gundlach's Hawk is larger; in flight, head extends well forward of wings; tail is rounded at tip.

RANGE. Throughout the Americas to northern Argentina; Bahamas, Cuba, Hispaniola, and Puerto Rico, wintering south to southern United States, Panama, and Greater Antilles. **Status**: Rare permanent resident on Cuba. Endangered. Migrants from North America are very rare winter residents. **Habitat**: Forests, mostly at moderate to high elevations. **Nesting**: Undescribed in Cuba. **Voice**: A high, repeated *kik*. **Food**: Almost entirely small birds.

GUNDLACH'S HAWK
GAVILÁN COLILARGO
Accipiter gundlachi   Pl. 12

DESCRIPTION. 19" (48 cm). Bluish gray above; white below with fine rufous barring. Older males attain gray underparts with very faint rufous barring. In flight, head extends well beyond wings. Wings short, narrow toward tips. Tail long, rounded, and banded; from above, narrowly tipped with white. Undertail coverts white as in Sharp-shinned Hawk. Immature brown above; heavily streaked with brown below. Soars occasionally.

SIMILAR SPECIES. (1) Sharp-shinned Hawk is smaller, with square tail. In flight, head does not extend far forward of wings. (2) Female Northern Harrier has white rump; in flight wings are held in a shallow V. (3) Broadwinged Hawk has shorter tail; wings are whitish below and black tipped.

RANGE. Endemic to Cuba, including Cayo Coco in the Archipiélago de Sabana-Camagüey. **Status**: Widely distributed, generally rare and local, although less rare at Zapata peninsula, Casilda, and Gibara. Vulnerable. **Habitat**: Semideciduous woods, swamp edges. **Nesting**: Mar–May. Nest placed at 10m or above in trees. Builds a flat twig platform, lined with grass. Lays two to four pale bluish-green eggs. **Voice**: Nest-defense call: a nasal, repeated *aah*; also a soft double whistle during the breeding season. In flight, a whistle. **Food**: Mainly birds, as large as parrots, rails, and pigeons.

COMMON BLACK-HAWK
GAVILÁN BATISTA; GAVILÁN CANGREJERO
Buteogallus anthracinus   Pl. 12

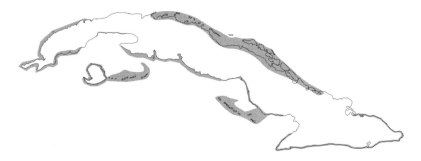

DESCRIPTION. 21ʺ (53 cm). Dark chocolate brown with broad, rounded wings and a white patch at base of primaries clearly visible in flight. Tail rather short with broad black and white bands. Base of bill and legs yellow to orange. Often seen perched. Immature streaked and paler, with distinct streaking on head and heavily banded tail. Flies with wings held exactly horizontal and shows upturned primary tips.

SIMILAR SPECIES. (1) Immature Broad-winged Hawk is smaller, with less distinct white markings on face and thinner legs. (2) Immature Red-tailed Hawk is larger, with white patch on chest. (3) Snail Kite has broad white tail base, red bill and legs, and long, finely hooked bill.

RANGE. Southwestern and south central United States through Central America to Guyana and Cuba. **Status**: Common permanent resident on Cuba, Isla de Pinos, and several larger cays. Cuban population may represent a distinct species, *B. gundlachii*. **Habitat**: Mangroves, coastal swamps. **Nesting**: Jan–Jul. Nests in trees, usually at moderate elevations. Lays three brown-spotted, creamy-olivaceous eggs. **Voice**: A repeated whistle suggesting the syllables *ba-tis-ta*, with stress on second syllable. **Food**: Mainly crabs, birds.

BROAD-WINGED HAWK
GAVILÁN BOBO; GUARAGUAO; GAVILÁN SONSO
Buteo platypterus   Pl. 12

DESCRIPTION. 15ʺ (38 cm). Dark brown above; barred below. Wings broad and slightly pointed at tips; underwing white with black tips. Immature has brown-streaked chest, with duller and more numerous tail bands.

SIMILAR SPECIES. (1) Gundlach's Hawk has longer tail and narrower wings with unmarked tips. (2) Immature Red-tailed Hawk is larger, with distinct white patch on chest. (3) Immature Common Black-Hawk is larger, with more distinct white markings on head, thicker legs, and, in flight, a white patch on underside of base of primaries is visible.

RANGE. Canada and eastern United States, wintering south to Bolivia and Brazil; Cuba, Puerto Rico, and from Antigua south to Grenada and Tobago. **Status**: Common permanent resident and rare migrant. Some migrants remain during the winter. **Habitat**: Dense, undisturbed semideciduous and pine forests. **Nesting**: Mar–Jul. Nests in trees. Builds a small and loose structure of twigs. Lays three white eggs, usually marked with brown. **Voice**: A two-toned whistle, the second note longer, *kee-deee*. **Food**: Mainly insects, birds, reptiles.

RED-TAILED HAWK
GAVILÁN DE MONTE; GAVILÁN POLLERO; GUARAGUAO
Buteo jamaicensis   Pl. 12

DESCRIPTION. 22″ (56 cm). Brown above; whitish below with large brown spots across the belly. Wings long, broad, and rounded with distinct dark bar and wrist patch on leading edge; primaries appear well separated when soaring. Rufous tail in adults; barred and brown tail in immatures. Soars frequently and high, sometimes in the same thermals as Turkey Vultures.

SIMILAR SPECIES. (1) Broad-winged Hawk is smaller, with boldly banded tail and black wing tips. (2) Immature Common Black-Hawk has streaked chest, thinly banded tail.

RANGE. Northern North America to Panama; northern Bahamas, Greater Antilles, and northern Lesser Antilles. **Status**: Common and widely distributed permanent resident in Cuba. Also on Cayo Coco in the Archipiélago de Sabana-Camagüey. **Habitat**: Low- to mid-elevation forests, including wooded swamps. **Nesting**: Jan–Jul. Nests usually in a tall tree. Builds a bulky structure of twigs with finer lining of stems and bark. Lays three white eggs, with blotches of buff and dark brown. **Voice**: A prolonged, harsh, slightly descending, hissing whistle, *keeeeer*, given in flight and more commonly during the breeding season. **Food**: Birds, immature hutias (endemic, arboreal rodents), knight anoles (large lizards), other vertebrates, insects.

CARACARAS AND FALCONS Falconidae

Medium-sized to large birds, with rather large heads and strong, hooked bills. Wings narrow and more or less pointed; tails long, generally tapering. Falcons are swift aerial hunters that capture prey such as dragonflies, lizards, and birds. Sexes usually alike. (**W**:63; **C**:4)

CRESTED CARACARA

CARAIRA

Caracara plancus   Pl. 13

DESCRIPTION. 22" (56 cm). Mostly black, grading to white toward the head, with a conspicuous black cap. Bare facial skin red to reddish orange. Wings rather wide and rounded, with white patches on outer wing. Tail white, finely barred, with a broad black subterminal band. Legs yellowish olive. Immature pale brown, with whitish throat and some distinct streaks on chest; gray legs. Rarely soars; commonly seen on the ground. In flight, has a distinctive deep wingbeat on the downstroke.

SIMILAR SPECIES. None

RANGE. Florida, northern Baja California, southern Arizona and Texas south through most of South America. Cuba. **Status**: Generally rare and local on Cuba and Isla de Pinos, but fairly common in the northern cays. **Habitat**: Pastures, palm-covered savannas, mangroves. **Nesting**: Mar–Dec. Nests are fairly large structures of sticks, plant stems, and weeds, on rather low palms and other trees, sometimes on large bromeliads. Lays two or three brown-spotted, chocolate-colored eggs. **Voice**: An alarm call *caracá–caracá*. **Food**: Carrion, slow-moving prey such as chicks and turtles.

AMERICAN KESTREL
CERNÍCALO

Falco sparverius   Pl. 13

DESCRIPTION. 10″ (25 cm). *Male*: bluish gray wings; unbarred tail with black subterminal band. *Female*: rufous wings; barred back and tail. Both sexes have rufous back and tail, two vertical black facial stripes on head, and false-eye spots on nape. The Cuban race has two color morphs, one with white chest and belly; the other with rufous chest and belly. Male is unmarked below; some may have spots on side. Female frequently has streaks on chest; others may have spots or barred belly. Immature streaked below. Wintering North American birds can be distinguished by their conspicuously streaked or spotted underparts suffused with buff. Often perches on high bare branches, tips of palm trees, telephone poles. Hovers over open areas with rapid wingbeats, searching for rodents, lizards, and insects.

SIMILAR SPECIES. Merlin has bluish gray (male) or dark brown (female) back, a single black facial line, and is strongly streaked below. Flight is usually much faster.

RANGE. North, Central, and South America; West Indies. **Status**: Common permanent and winter resident in Cuba, Isla de Pinos, and some cays. **Habitat**: Open country, cultivated lands, often around built-up areas. **Nesting**: Dec–May. Nests in abandoned woodpecker holes, topless dead palms, natural cavities in trees. Lays three or four brown-spotted, cream-colored eggs. **Voice**: A high-pitched, repeated *killy-killy-killy*. **Food**:

Lizards (*Ameiva, Anolis, Leiocephalus*), large insects such as grasshoppers and dragonflies, mice, small birds. Wintering birds have been seen preying on bats.

MERLIN
HALCONCITO DE PALOMAS; HALCONCITO; FALCÓN

Falco columbarius   Pl. 13

Description. 12" (30 cm). *Male*: dark bluish gray above. *Female*: brown above with gray uppertail coverts. Both sexes buffy below with brown streaks, and poorly defined vertical stripe below eye. Tail boldly banded in light gray and black, with a white tip. Flight very fast. Immature similar to female, but uniformly brown above. Almost invariably appears slightly heavier, darker, and faster than American Kestrel.
Similar species. (1) American Kestrel has rufous tail and back. (2) Peregrine Falcon is larger, with well-defined facial markings.
Range. Northern Hemisphere, wintering south to West Indies and South America. **Status**: Uncommon winter resident and transient in Cuba, Isla de Pinos, and some larger cays, sometimes locally common (2 Sep–2 May). **Habitat**: Open fields, low coastal vegetation. **Voice**: A whickering trill. **Food**: Small birds, rodents, large insects.

PEREGRINE FALCON
HALCÓN PEREGRINO; FALCÓN; HALCÓN DE PATOS

Falco peregrinus   Pl. 13

Description. 18" (46). Dark brown to slaty gray above with black head and whitish breast and belly, barred with brown. A distinct black vertical facial stripe or "moustache." Long, tapered tail, faintly barred. Immature is brown above, with chest and belly streaked with brown. Males are markedly smaller than females.
Similar species. Merlin is smaller, with less striking facial markings.
Range. Worldwide; North American populations winter to South America and West Indies. **Status**: Uncommon winter resident and transient that may also breed in Cuba, Isla de Pinos, and some northern cays (6 Oct–17 Apr). **Habitat**: Swamps, marshes, rice fields, flooded savannas, cities. **Voice**: A harsh *kak-kak-kak-kak*; also a series of discordant, interrupted notes. **Food**: Mainly birds, especially aquatic ones; in urban areas mostly pigeons; occasionally bats.

PHEASANTS, QUAILS, AND GUINEAFOWL Phasianidae

These heavy bodied birds are fast walkers with short, rounded wings; they rarely fly. Bill short. Food is mainly seeds, gathered from the ground, where vegetation is rather dense. Usually observed flying low for short distances, after being disturbed. In general, sexes differ. (**W**:177; **C**:3)

RING-NECKED PHEASANT
FAISÁN DE COLLAR

Phasianus colchicus   Pl. 51

DESCRIPTION. Male: 33" (84 cm); Female: 21" (53 cm). *Male*: large, multi-colored, and iridescent. Green to purple head, with conspicuous bare red skin on face and usually a white neck ring. Long and pointed tail. *Female*: smaller, mottled brown, with shorter tail. Both sexes have short rounded wings and when flushed arise with a sudden burst of loud whirring.
SIMILAR SPECIES. None.
RANGE. Asia. Widely introduced in Europe, Hawaiian Islands, New Zealand, and North America. **Status**: Introduced with very limited success at Guanahacabibes and Topes de Collantes; common at Los Indios (Isla de Pinos). **Habitat**: Grassy open country. **Nesting**: Undescribed in Cuba. **Voice**: Male: a loud crowing *haa-haak*, followed by a whirr of wings. Both sexes emit croaking alarm notes. **Food**: Omnivorous.

NORTHERN BOBWHITE
CODORNIZ

Colinus virginianus   Pl. 13

DESCRIPTION. 9" (23 cm). Chunky. *Male*: reddish brown and mottled, with white eyebrow and throat and a large black patch on breast. *Female*: brown above with whitish barred belly and buffy throat. Both sexes have reddish brown sides and flanks, and a short gray tail. Juvenile similar to adult female, but duller. Gathers in small flocks.
SIMILAR SPECIES. None.
RANGE. Southern Ontario, central and eastern United States to Guatemala. Cuba. Introduced in Hispaniola, Puerto Rico, St. Croix, and Andros and New Providence in the Bahamas. **Status**: Common permanent resident, possibly introduced. **Habitat**: Savannas, pastures with nearby dense cover. **Nesting**: Apr–Jul. Nest is a shallow depression lined with grass, with a side entrance; built among grass. Lays up to 18 dull-white eggs. **Voice**: A bisyl-labic whistle, the second part louder and more extended: *bob-WHITE*. **Food**: Seeds, fruits, shoots, insects.

HELMETED GUINEAFOWL

GALLINA DE GUINEA; GUINEA; GUINEO

Numida meleagris   Pl. 13

DESCRIPTION. 22" (56 cm). Grayish black, with plumage entirely and very finely spotted with white. Wholly or partially white birds are occasionally encountered, usually around farms with other domestic birds. Head and neck naked. An excellent runner that only rarely takes flight.

SIMILAR SPECIES. None.

RANGE. Africa. **Status**: Introduced during the slave trade and established throughout Cuba near farms and towns. **Habitat**: Common in pastures, savannas, grassy fields with low bushes. **Nesting**: Nests on the ground. Lays up to 16 pale-brown eggs. **Voice**: A loud, grating *cherrrr* or *kek-kek-kek-krrrr* alarm call. Also a loud *Pas-cual, Pas-cual.* Breeding female makes a repeated *pittoo* call. **Food**: Seeds, insects, mollusks, shoots.

RAILS, GALLINULES, AND COOTS Rallidae

Small to medium-sized, more or less aquatic birds with long legs and toes, rather heavy-looking bodies, and short rounded wings. Gallinules and coots swim well and spend significant periods of time in the water. Seeds and grain are included in their diet, as are aquatic invertebrates. Rails, on the other hand, swim only occasionally and are extremely secretive and almost strictly carnivorous. They take flight reluctantly, their legs dangling beneath them. All are good runners. Nest is usually a bulky cup of dead vegetation. Newly hatched chicks are mostly dark brown to black. Sexes alike. (**W**:142; **C**:12)

BLACK RAIL

GALLINUELITA PRIETA

Laterallus jamaicensis   Pl. 14

DESCRIPTION. 5.5" (14 cm). Very small. Very dark slaty gray; back and wings covered with small white spots; nape deep chestnut, flanks barred with white. Short and black bill. Juvenile similar to adults, but duller. Strictly nocturnal.

SIMILAR SPECIES. Very young rails of other species have entirely black plumage.

RANGE. Central and eastern United States and California, Belize, Peru, Chile, and western Argentina; Cuba, Jamaica, and Puerto Rico. **Status**: Rare winter resident and transient, possible permanent resident in Cuba (3 Jul–Apr). **Habitat**: Coastal marshes, shores of natural and man-made freshwater bodies. **Nesting**: Undescribed in Cuba. A small and deep cup of grasses and sedges, with overhanging adjacent grasses or plant stems being pulled and woven together to form a concealing canopy. **Voice**: Male, *kee-kee-keer*; female, *whoo-who.* **Food**: Mostly insects, crustaceans, seeds.

CLAPPER RAIL

GALLINUELA DE MANGLAR; GALLINUELA DE AGUA SALADA

Rallus longirostris   Pl. 15

DESCRIPTION. 14.5″ (37 cm). Blackish brown above with ashy feather margins; lower belly rather faintly barred with white; cheeks gray. Legs and bill long. Tail very short, often cocked. Juvenile paler below.

SIMILAR SPECIES. (1) The more warmly toned King Rail has black on central back feathers edged with tawny, rusty wing coverts, and more distinct belly barring. (2) Virginia Rail is much smaller, with reddish bill and gray cheek contrasting with rusty brown chest.

RANGE. Coasts of United States to southeastern Brazil. West Indies. **Status:** Common permanent resident in Cuba, Isla de Pinos, and many cays. Winter residents have been reported. **Habitat:** Coastal vegetation, especially mangrove. **Nesting:** Apr–Jun. Nest is a bulky cup elevated on a firm bank among mangrove roots. Lays five to nine cream-colored eggs, spotted with dark red. **Voice:** A long series of *kek* notes rising and falling. **Food:** Crabs, small mollusks, fishes, aquatic insects, plants.

KING RAIL

GALLINUELA DE AGUA DULCE; GALLINUELA; MARTINETE;
GALLINETA

Rallus elegans   Pl. 15

DESCRIPTION. 15" (38 cm). Blackish brown above with tawny feather margins; lower belly boldly barred; wings, chest, and cheeks rusty brown. Legs and bill long; tail very short, frequently cocked. Juvenile is paler below.

SIMILAR SPECIES. (1) Less richly colored Clapper Rail has black back feathers with ashy margins and rather faintly barred belly. (2) Virginia Rail is much smaller with gray cheeks.

RANGE. Central and eastern United States, Mexico. **Status**: Common permanent resident and migrant in Cuba and Isla de Pinos. **Habitat**: Freshwater swamps, marshes, lagoons, rice fields. **Nesting**: May–Dec. Nests on the ground in a grass tussock or waterside vegetation. Builds a cup of grasses with growing stems pulled over to form a canopy. Lays up to nine cream-colored eggs spotted with reddish brown. **Voice**: A repeated *kek-kek-kek, trr-trr-trrrr*. **Food**: Seeds, small fruits, insects, crustaceans, frogs, mollusks.

VIRGINIA RAIL

GALLINUELA DE VIRGINIA

Rallus limicola   Pl. 14

DESCRIPTION. 10" (23–28 cm). Brown, mottled above, with flanks strongly barred with white. Wings and chest rusty brown; cheeks gray. Legs and bill long, usually reddish. Juvenile very dark, with mostly black underparts.

SIMILAR SPECIES. (1) King Rail is much larger with brown cheeks. (2) Clapper Rail is larger and grayer, with dull barring on belly.

RANGE. North America and Mexico; Colombia, and Peru. North American populations winter south to Guatemala. **Status**: Vagrant. Two records: 16 Oct (1958); Dec (1995). Gibara, Zapata peninsula. **Habitat**: Brackish and saltwater estuaries, densely vegetated freshwater canals. **Voice**: A descending series of grunts; also a *ticket-ticket-ticket-trr*, or *kek-kidick-kidick*. **Food**: Snails, earthworms, frogs.

SORA

GALLINUELA OSCURA; GALLINUELA CHICA; GALLINUELITA

Porzana carolina   Pl. 14

DESCRIPTION. 9" (23 cm). Brown; belly barred with black and white. Bill short, stocky, and yellow to greenish yellow. In breeding plumage, face, throat, and chest are black, fading to gray on sides. In winter plumage, throat and chest are mostly gray. Female duller. Juvenile has buff brown face, throat, and chest.

SIMILAR SPECIES. Yellow-breasted Crake resembles juvenile Sora, but is much smaller with black cap and line through eye; bill finer and olive black.

RANGE. Breeds in North America, wintering south to Peru; West Indies. **Status**: Common winter resident and transient in Cuba and Isla de Pinos (9 Sep–14 May). **Habitat**: Salt marshes, freshwater swamps. **Voice**: A thin whistle, *too-wee*, also a descending whinny. **Food**: Grass seeds, small mollusks, worms, insects.

YELLOW-BREASTED CRAKE
GALLINUELITA

Porzana flaviventer   Pl. 14

DESCRIPTION. 5.5" (14 cm). Small, and buffy brown above, streaked with black and white; belly barred with gray and white. Chest, neck, and head washed with ochre, with black crown and stripe through eye. Short, olive bill; yellowish legs.

SIMILAR SPECIES. Juvenile Sora resembles Yellow-breasted Crake, but is much larger with plain face; Deep-based yellowish bill.

RANGE. Southern Mexico to Argentina; Greater Antilles. **Status**: Very rare permanent resident in Cuba and Isla de Pinos. **Habitat**: Well-vegetated shores of lakes, lagoons, and swamps, especially where water hyacinths abound. **Nesting**: Apr–Oct. Undescribed in Cuba. Known elsewhere to nest among aquatic plants, laying four or five spotted pale-cream eggs. **Voice**: A repeated thin, high-pitched *peep*. **Food**: Plants, aquatic and terrestrial invertebrates, and their eggs.

ZAPATA RAIL
GALLINUELA DE SANTO TOMÁS

Cyanolimnas cerverai   Pl. 14

DESCRIPTION. 11.5" (29 cm). Olive brown above with slaty gray under-parts; white throat, buffy undertail coverts bordered with white; flanks and lower belly barred with white in some individuals. Green bill with red base; red legs. Juvenile duller, lacks red on bill, and has olivaceous legs. The phantom of Cuba's avifauna, occasionally heard but very rarely seen. Flight is very weak.

SIMILAR SPECIES. Spotted Rail is black, extensively marked with white.

RANGE. Endemic to Zapata peninsula, Cuba. **Status**: Endangered. Several sightings from north of Santo Tomás, and three sightings from Laguna del Tesoro. Recently (1998) found in Peralta and Nato de Jicarita, Zapata peninsula. Known as fossil from Pinar del Río and Isla de Pinos. **Habitat**: Sawgrass marshes. **Nesting**: Nov–Jan? Nest and eggs undescribed. Breeding biology data, cited in most previous relevant literature, is ambiguous and does not certainly refer to this species. Therefore, we prefer to disregard it. **Voice**: A repeated *kuck*, a repeated, low-pitched *cutucutu*, reminiscent of the call of Bare-legged Owl, and also a guttural grunt reminiscent of Spotted Rail. **Food**: Unknown. Possibly tadpoles and aquatic invertebrates.

SPOTTED RAIL

GALLINUELA ESCRIBANO

Pardirallus maculatus   Pl. 14

DESCRIPTION. 11" (28 cm). Black with white spots above and on chest; belly barred with white. Undertail coverts white; wings tawny. Bill olive with red base to lower mandible and bright green base to upper mandible; red eye; legs reddish. Juvenile brown, with duller spots, and olivaceous legs.

SIMILAR SPECIES. Zapata Rail is olive brown above and uniform grey below, with white throat.

RANGE. Central America to Argentina; Cuba, Dominican Republic, and perhaps formerly in Jamaica. **Status**: An uncommon permanent resident in western and central Cuba and Isla de Pinos. **Habitat**: Grassy marshes such as found north of Santo Tomás in Zapata peninsula and Laguna El Corojal; also lagoons with dense vegetation and rice fields. **Nesting**: Jul–Oct. Undescribed in Cuba; elsewhere among grass, laying three to seven spotted eggs. **Voice**: A high-pitched guttural grunt, more like a pig than a bird; also a fast, repeated *tuk*. **Food**: Invertebrates, small frogs.

PURPLE GALLINULE
GALLARETA AZUL
Porphyrula martinica   Pl. 15

DESCRIPTION. 13″ (33 cm). Purple blue, with iridescent green back and wings. Red bill, tipped with yellow; a bright pale-blue frontal shield; yellow legs. Juvenile entirely buffy brown, with a green sheen on wings and back; bill mostly dark olive. Both adult and juvenile have white undertail coverts. Chick blackish, with yellow legs; bill yellow, with white tip and black ring at center. Surprisingly adept at swimming, walking over aquatic vegetation, climbing over bushes, and flying. Cocks tail frequently while walking.

SIMILAR SPECIES. (1) Juvenile Common Moorhen is dark gray, with white stripes on sides. (2) Juvenile American and (3) Caribbean Coots are dark gray with white bill.

RANGE. Southeastern United States and Mexico south to Argentina; West Indies. **Status**: Common permanent resident in Cuba and Isla de Pinos; also a winter resident. **Habitat**: Swamps, lagoons, canals with dense aquatic vegetation, rice fields. **Nesting**: Feb–Sep. Builds a bulky cup of dead or green plant material, well above the water. Lays up to 12 eggs, spotted with reddish brown. **Voice**: A series of cackling interspersed with much deeper sounds: *Kr-rruk, kek, kek, kek*. **Food**: Mostly seeds, snails, aquatic insects, snails, frogs.

COMMON MOORHEN
GALLARETA DE PICO ROJO; GALLARETA DE PICO COLORADO
Gallinula chloropus   Pl. 15

DESCRIPTION. 13″ (33 cm). Slaty gray; brownish on back and wings. Bill red except for yellow tip; frontal shield red; legs green. Adult and juvenile have white stripes on sides; undertail coverts white on sides, black in cen-

ter. Juvenile is gray, with brownish back; bill and legs dusky. Chick black-ish, with a greenish cast above, browner below; bluish skin on crown; or-ange red bill with yellow tip and dark olive legs.

SIMILAR SPECIES. (1) Juvenile American and (2) Caribbean Coots are en-tirely dark gray, with white bill. (3) Juvenile Purple Gallinule is buffy brown, with a green sheen on back and all-white undertail coverts; cocks tail frequently when standing.

RANGE. Southeastern Canada, southwestern and eastern United States to Chile and northern Argentina; Europe, Africa, and Asia; West Indies. **Sta-tus**: Common permanent resident in Cuba and Isla de Pinos; also a winter resident. **Habitat**: Freshwater lakes and reservoirs; brackish ponds. **Nest-ing**: May–Dec. Builds a bulky platform of dead plant material and debris near or over the water, laying up to nine grayish-white, reddish-brown-spotted eggs. **Voice**: Clucks, cackles, and harsh cries. *Kr-r-ru, kruh, kruh, kruh.* **Food**: Aquatic vegetation, mollusks, worms, fruit.

AMERICAN COOT

GALLARETA DE PICO BLANCO; GALLARETA AMERICANA
Fulica americana

Pl. 15

DESCRIPTION. 15″ (38 cm). Dark slaty gray; black toward the head. Bill and frontal shield white, except for apex of shield, which is maroon. Legs green, with lobed toes. Both adult and juvenile have white undertail coverts with black centers. Juvenile paler, with white bill. Chick blackish, with orange head, neck, and throat; blue skin over eye; reddish, white-tipped bill; and gray legs. May form loose social groups or dense flocks. Dives for food.

SIMILAR SPECIES. (1) Caribbean Coot has all-white frontal shield. (2) Juvenile Purple Gallinule is buffy brown, with a green sheen on back and wings; undertail coverts are entirely white. (3) Juvenile Common Moorhen has thin white stripes on sides and brownish bill.

RANGE. North and Central America wintering south to Colombia; Bahamas, Jamaica, Hispaniola, and Grand Cayman. **Status**: Abundant permanent and winter resident in Cuba, Isla de Pinos, and several cays. **Habitat**: Both natural and man-made freshwater and brackish water. **Nesting**: May–Dec. Builds a bulky cup of dead leaves and stems of waterside plants. Nests near or over water, laying up to 12 eggs heavily spotted with dark brown and black. **Voice**: A typical *took*, repeated at irregular intervals, and varied clucks and cackles. **Food**: Aquatic vegetation, snails, tadpoles, worms.

CARIBBEAN COOT
GALLARETA DEL CARIBE
Fulica caribaea Pl. 15

DESCRIPTION. 15" (38 cm). Similar to American Coot, but shield entirely white and broader, sometimes tinged with yellow. The status of this species is uncertain, possibly being a freely interbreeding morph of American Coot.

SIMILAR SPECIES. (1) American Coot has dark maroon apex to frontal shield. (2) Juvenile Purple Gallinule is buffy brown, with a green sheen on back and wings; undertail coverts are entirely white. (3) Juvenile Common Moorhen has thin white stripes on sides.

RANGE. Northwest Venezuela; most of the Antilles and Grand Cayman. **Status**: No specimens collected for over 50 years in Cuba. Early specimens may represent vagrants, or perhaps remnants of a relict population. Has been observed in some lagoons of Pinar del Río province, such as La Deseada. **Habitat**: Ponds and lakes. **Voice**: Undescribed in Cuba, but known to be similar to American Coot.

LIMPKINS Aramidae

Monotypic family. Sexes alike. (**W**:1)

LIMPKIN
GUAREAO
Aramus guarauna   Pl. 15

DESCRIPTION. 27" (68 cm). A large solitary wading bird, with long legs, neck, and slightly decurved bill; brown, speckled with white from mid-body to head. Juvenile paler and more spotted than adult. Has a hunched look in flight and the upstroke is accomplished with a distinctive flick. Wings barely come above the horizontal. On the ground, flicks tail regularly. Usually found near the edge of water.

SIMILAR SPECIES. (1) Glossy Ibis lacks the bold speckling and its bill is more slender and more markedly decurved. (2) Immature White Ibis has white chest and belly.

RANGE. Southeastern United States to northern Argentina; Greater Antilles. **Status**: Common permanent resident in Cuba, Isla de Pinos, and some larger cays. **Habitat**: Freshwater marshes, swamps, ponds, humid forests. **Nesting**: May–Jan. Nest is near water, on the ground or low in trees. Nest is a loosely constructed saucer of plant stems, dead leaves, dead vines, and various plant materials. Lays three to five creamy-buff eggs, heavily spotted and blotched with brown, pale gray, and lilac. **Voice**: A wild and far-carrying *karrao*, and a short knocking note. Highly vocal, especially at dusk and at night. **Food**: Mainly apple snails (*Pomacea*); also muddy shore invertebrates.

CRANES Gruidae

Very large birds, with long legs and neck, and a heavy pointed bill. Usually gather in small flocks that feed by day in open terrain with low grass. Sexes alike. (**W**:15; **C**:1)

SANDHILL CRANE
GRULLA
Grus canadensis   Pl. 15

DESCRIPTION. 40" (102 cm). Gray, with a red featherless patch on fore-head. At rest, the elongated, bushy tertials cover the rump area. Flies with neck extended, the wings never going far below the horizontal; may glide for short distances. Immature has neck and entire head brownish; mot-tled above.

SIMILAR SPECIES. Great Blue Heron has yellowish bill and broad black stripes on head; neck is retracted in flight.

RANGE. Northeastern Siberia, North America, Cuba. American popula-tions winter south to Mexico. **Status**: An endangered permanent resident in Cuba, of restricted distribution: Santo Tomás, San Lázaro (Zapata peninsula), northern and southwestern Isla de Pinos (Los Indios), clear-ings in pine forests of Viñales valley, areas north of Itabo in Matanzas province, and several areas of Ciego de Ávila and Camagüey provinces. **Habitat**: Savannas, marshes, grassy fields near pine forests. **Nesting**: Feb–Jul. Large heap of plant material gathered from around site, with slight central hollow. Lays two grayish-green eggs, spotted and blotched with brown or darker reddish brown. **Voice**: A loud rolling *kurooo*. **Food**: Omnivorous.

PLOVERS Charadriidae

Small to medium-sized shorebirds with rather large head, long legs, and a short, straight bill. Quite similar in shape, these birds differ mainly in size, habitat preference, and the pattern of black and white markings below. Roaming beaches, savannas, and pastures, they characteristically run a few steps and pause motionless before pecking at a food item or continuing. All lay eggs on the ground. Many are long-distance migrants. (**W**:89; **C**:7)

BLACK-BELLIED PLOVER

PLUVIAL CABEZÓN; TÍTERE

Pluvialis squatarola Pl. 16

DESCRIPTION. 12″ (30 cm). Winter plumage brownish gray above, whitish below, with lightly streaked chest. Eyebrow whitish. In flight, a conspicuous black patch is visible at base of underwing. Bill and legs black. Breeding plumage is black below with white undertail coverts; back scaled black and white; forehead and crown white. In flight, uppertail coverts and median wing stripe are whitish. Juvenile similar to winter plumage adults, slightly more streaked below.

SIMILAR SPECIES. American Golden-Plover is smaller and has dark uppertail coverts. Base of underwing is unmarked; bill is finer.

RANGE. Breeds in high Arctic, wintering south to Chile and Argentina; West Indies. **Status**: Common winter resident and transient in Cuba, Isla de Pinos, and some cays (18 Jul–May). **Habitat**: Muddy and sandy flats near the sea, lagoons, marshes. **Voice**: A sweet three-syllable whistle, *pee-u-wee*. **Food**: Worms, especially polychaetes, clams, other marine invertebrates.

AMERICAN GOLDEN-PLOVER

PLUVIAL DORADO

Pluvialis dominica Pl. 16

DESCRIPTION. 10.5″ (27 cm). Winter plumage grayish golden above, white below, with lightly streaked on chest. Eyebrow white. Bill and legs black. Summer plumage is black below, dark above mottled with golden brown, with a distinct S-shaped white stripe from forehead to side of lower chest. Wing unmarked. Juvenile is similar to winter adult, but with slightly browner side and belly.

SIMILAR SPECIES. Black-bellied Plover is larger; winter plumage is gray above; in flight, a conspicuous black patch shows at base of underwing; uppertail coverts are white; bill is heavier.

RANGE. Northern North America, wintering south to northern Chile and Argentina. **Status**: Very rare transient in Cuba (26 Jul–4 Dec; 4 Feb–Apr). **Habitat**: Plowed fields, coasts, shores of lagoons. **Voice**: A rich, trebled *queedle* given in flight. **Food**: Worms, spiders, rarely seeds and berries.

SNOWY PLOVER

FRAILECILLO BLANCO

Charadrius alexandrinus Pls. 16, 19

DESCRIPTION. 6″ (15 cm). Breeding plumage very pale sandy gray above, white below. Bill thin, black; legs dark gray. Patches behind eye, on side of breast, and on crown are black in male, dark in female. Dark or black

patches become pale in winter adults. Juvenile has pale brown patch on sides of breast, often indistinct; grayish brown back; and buff-tipped feathers.

SIMILAR SPECIES. (1) Semipalmated and (2) Wilson's Plovers are brown above. (3) Piping Plover has bright orange legs.

RANGE. Breeds locally from western and southern United States, northeastern Mexico, southern Bahamas, Greater Antilles, northern Lesser Antilles. In South America, Chile and Peru. Eurasia and North Africa. North American populations winter on coasts of southern and western United States south to Panama, Greater Antilles, and in the breeding range in South America. **Status**: An endangered permanent resident, known to breed near Guantánamo and Cayo Sabinal; probably also in Salinas de Bidos, Itabo. **Habitat**: Sandy beaches. **Nesting**: Apr–Aug. Nests directly on sand, laying three black and pale-gray-spotted buff or sandy eggs. Markings, at times, concentrated toward the larger end. **Voice**: A low *krut*, and a whistled *ca-wee*. **Food**: Insects, worms, mollusks, fish.

WILSON'S PLOVER
TÍTERE PLAYERO; PUTILLA DE PLAYA; FRAILECILLO
Charadrius wilsonia   Pls. 16, 19

DESCRIPTION. 8" (20 cm). Breeding adult brown above, white below. Bill black, rather long, and thick; legs pale pink. Single, wide breast band and forehead black in male, brown in female, pale brown in winter adults. Juvenile resembles female, with very pale or incomplete breast band and the back appearing scaled.

SIMILAR SPECIES. (1) Semipalmated Plover is smaller, with narrow breast band, short bill, and orange or yellow legs. (2) Piping and (3) Snowy Plovers are smaller and pale sandy gray above.

RANGE. Coastal southeastern United States south to Brazil; West Indies. **Status**: Common summer resident and transient in Cuba, Isla de Pinos, and some cays (19 Feb–14 Nov). A winter resident in some places (Cayo Coco). **Habitat**: Sandy shores and open mudflats. **Nesting**: Apr–Jul. Nest is a shallow scrape, unlined or lined with fragments of shells and pebbles. Lays three creamy eggs, heavily spotted, scrawled, and blotched with black and pale gray. **Voice**: A sharp *wheet*. **Food**: Insects, crustaceans, worms, small snails, insect larvae.

SEMIPALMATED PLOVER
FRAILECILLO SEMIPALMEADO
Charadrius semipalmatus   Pl. 16

DESCRIPTION. 7" (18 cm). Brown above, white below; bill very short. Winter adult has a single brown narrow breast band and brown forehead; bill dull yellow at base. Adults in breeding plumage have black breast band, orange yellow bill with black tip, and orange yellow legs. Juvenile resembles winter adult, with dark bill and legs and a slightly scaled appearance above.

SIMILAR SPECIES. (1) Piping and (2) Snowy Plovers are pale sandy gray above, with incomplete breast band. (3) Wilson's Plover has a thicker breast band, larger bill, and pale pink legs.

RANGE. Northern North America, wintering on both coasts south to Chile and Argentina; West Indies. **Status**: Common winter resident and transient in Cuba, Isla de Pinos, and some cays (26 Sep–Apr). **Habitat**: Muddy terrain, either dry, wet, or flooded, with freshwater or salt water; sandy beaches. **Voice**: A whistled *chu-wee*. **Food**: Mollusks, crustaceans, worms, insects.

PIPING PLOVER
FRAILECILLO SILBADOR
Charadrius melodus   Pls. 16, 19

DESCRIPTION. 7" (18 cm). Pale sandy gray above, white below. Bill very short, orange at base; legs bright orange. Though breeding birds show a variably complete or incomplete black breast band, this is paler or almost lacking in birds visiting Cuba in winter. In breeding plumage, forehead is black. Juvenile resembles winter adult, with buff-tipped back feathers, dull orange legs, breast band absent, and forehead unmarked.

SIMILAR SPECIES. (1) Semipalmated and (2) Wilson's Plovers are brown above. (3) Snowy Plover has thin black bill and dark gray legs.

RANGE. Northern and eastern North America, wintering to southeastern United States; Bahamas and Greater Antilles. **Status**: Vulnerable. An uncommon, but local winter resident and transient to Cuba; more common on central northern cays (Aug–Apr). **Habitat**: Sandy shores, both freshwater and salt water. **Voice**: A whistled *peep-lo*. **Food**: Marine worms, mollusks, crustaceans.

KILLDEER

TÍTERE SABANERO; FRAILECILLO GRITÓN; TINGUILICHO

Charadrius vociferus   Pls. 16, 19

DESCRIPTION. 10″ (25 cm). Brown above, white below, with two bold black breast bands. An extensive bronze-orange rump is conspicuous in flight. Bill rather long, black; legs dusky. Juvenile during first weeks has only one breast band. Very active at night.

RANGE. Northern North America south to central Mexico; in South America along the coast of Peru and northwestern Chile; Greater Antilles. **Status**: Common in Cuba, Isla de Pinos, and some cays, both as permanent and winter resident. Also a transient. **Habitat**: Savannas, seashores, and open country, often far from water. **Nesting**: Mar–Jul. Nests on the ground. A shallow scrape, unlined or lined with nearby materials such as pebbles, woodchips, plant fragments. Lays three or four pale-buff eggs with spots, blotches, or scrawls of black and pale gray. **Voice**: A particularly noisy species, giving a loud, repeated *kill-deeah*. **Food**: Insects, weed seeds.

OYSTERCATCHERS Haematopodidae

Medium-sized birds of rocky or sandy shores. Bills are red or orange, long, straight, and laterally compressed. Feed on shellfish and other invertebrates. Neck and tail are short. Sexes alike. (**W**:11; **C**:1)

AMERICAN OYSTERCATCHER

OSTRERO

Haematopus palliatus   Pl. 16

DESCRIPTION. 18" (46 cm). Dark brown above; white below; black head
and neck. Median wing stripe and tail patch white. Bill bright red; legs
pink. Juvenile has brownish head and neck, dark bill, and gray legs.

SIMILAR SPECIES. None.

RANGE. Southwestern and eastern North America south to Chile and Ar-
gentina; West Indies. **Status**: Very rare winter resident, from both United
States and Bahamian populations. Recently observed at Cayo Coco and
Cayo Paredón Grande (Sep–Oct); also once in Zapata peninsula: 25 Jan.
Habitat: Beaches. **Voice**: A loud, screaming *wheep* or *wheet*. **Food**: Shell
fish, marine worms, fish.

STILTS AND AVOCETS Recurvirostridae

A small group of medium-sized to large waders with very long slender legs
and bill, and rather long neck. Head rather small and rounded. Plumage is
boldly patterned in black, browns and white. Usually seen in small flocks.
They feed on aquatic insects, mollusks, crabs and marine worms. Permanent
resident and migratory populations are known. Some species are very noisy.
Sexes alike. (**W**:13; **C**:2)

BLACK-NECKED STILT

CACHIPORRA; SOLDADO; MIGUELETE

Himantopus mexicanus   Pl. 16

DESCRIPTION. 14" (36 cm). Very slender and long-legged. *Male*: black
above, except rump and tail; white below; black wing linings. Bill black,
needlelike; legs pinkish red. *Female*: Similar to male, but brownish above.
Juvenile resembles female, with buffy edgings and paler legs.

SIMILAR SPECIES. None.

RANGE. United States to southern South America; West Indies. **Status**: Abundant permanent and winter resident; also a transient in Cuba, Isla de Pinos, and some cays. **Habitat**: Natural or man-made bodies of water, either fresh or salt. **Nesting**: Apr–Jun. Nests on the ground, in a shallow hollow, with a variable amount of nest material. Lays three or four pale brownish-buff eggs, usually marked with small black spots and blotches. Sometimes grayish markings. **Voice**: A loud, repeated *kek*. **Food**: Crayfish, brine shrimp, snails, tadpoles, fish, seeds.

AMERICAN AVOCET
AVOCETA

Recurvirostra americana   Pl. 16

DESCRIPTION. 18" (46 cm). Slim; white, with conspicuous black and white pattern on back and wings. Bill very slender and upturned, more so in females; legs long, grayish blue. Head and neck cinnamon in breeding plumage, but pale gray in birds visiting Cuba in winter. Juvenile resembles summer adults, with cinnamon restricted to dorsal part of head and neck.
SIMILAR SPECIES. None.

RANGE. Breeds in western and central North America, northwestern and central Mexico, wintering south to Costa Rica and southern Florida. **Status**: Vagrant. Five records: Gundlach, last century; 16 Nov (1963); 9 Jun (1995); Oct (1997); 23 Jan (1999). Casilda, El Jíbaro, Guanahacabibes; Cayo Coco. **Habitat**: Lagoons and marshes. **Voice**: A loud piping *wheeep*. **Food**: Mostly crustaceans, insects, seeds.

JACANAS Jacanidae

Medium-sized birds with small head and bill and rather long neck. Toes are extremely long, specialized for walking, foraging and nesting on floating vegetation. Food is mainly small crustaceans, worms, mollusks. Sexes alike. (**W**:8; **C**:1)

NORTHERN JACANA
GALLITO DE RÍO

Jacana spinosa   Pl. 14

DESCRIPTION. 9" (23 cm). Cinnamon brown, darkening to black toward neck and head. Bill, frontal shield, and flight feathers are bright yellow. Wing has a sharp, forward-pointing spur on "wrist." Female brighter than male. Immature is brownish above, white below, with a thin white stripe above eye, and the yellow wing pattern of adults. Flight is slow and low, with bursts of shallow flaps and periods of gliding. Raises wings over the body briefly when alighting.

SIMILAR SPECIES. None.

RANGE. Southwestern United States to Panama; Cuba, Hispaniola, and Jamaica. **Status**: Common permanent resident on Cuba and Isla de Pinos. **Habitat**: Marshes, lagoons, swamps. **Nesting**: Mar–Sep. Builds simple nest atop aquatic vegetation. Females often pair with several males, each of which builds a nest and incubates the three or four eggs, which are golden buff and heavily scrawled and scribbled with irregular patterns of thin black lines. **Voice**: Noisy, producing a peculiar loud cackling. **Food**: Small invertebrates and some fish.

SANDPIPERS AND PHALAROPES Scolopacidae

Small to large waders with thin and sometimes very long bills, and long and slender legs. Plumage is usually dull gray, brown, or buff, and streaked or mottled. Most species breed in Arctic regions, many adults migrating earlier in fall than juveniles. Usually found in flocks on or near beaches, as well as on the shores of lagoons or small muddy ponds. Sandpipers generally feed by probing into the water, mud, or sand, taking arthropods, worms, and mollusks. Sexes similar. (**W**:88; **C**:27)

GREATER YELLOWLEGS

ZARAPICO PATIAMARILLO GRANDE

Tringa melanoleuca   Pls. 17, 19

DESCRIPTION. 14″ (36 cm). Winter plumage mottled grayish brown above, white below; throat and chest faintly streaked. Long, straight bill in some birds, but slightly upturned in most; yellow legs. Breeding plumage dark brown above marked with black and white; throat, chest, and underparts heavily streaked. Both plumages have finely barred white tail, conspicuous in flight. Juvenile similar to breeding adults, but duller, with buffy spots. Frequently nods or teeters.

SIMILAR SPECIES. (1) Lesser Yellowlegs is smaller, with thinner and shorter straight bill. (2) Willet has thick straight bill, gray legs; in flight, shows very striking white wing stripe.

RANGE. Northern North America, wintering south to southern Argentina; West Indies. **Status**: Common winter resident and transient in Cuba, Isla de Pinos, and several cays (Aug–Apr). **Habitat**: Flooded fields, lagoons, marshes, seashores. **Voice**: A loud, ringing *tew -tew-tew* or *dear-dear-dear*. **Food**: Small fish, insects, snails, worms, tadpoles.

LESSER YELLOWLEGS

ZARAPICO PATIAMARILLO CHICO

Tringa flavipes   Pls. 17, 19

DESCRIPTION. 10.5" (27 cm). Winter plumage mottled grayish brown above, white below; faintly streaked. Breeding plumage more streaked on throat and chest, marked with black and white above. Moderately long, straight bill; yellow legs. Finely barred white tail is conspicuous in flight. Juvenile resembles breeding adults, with breast indistinctly streaked and washed with gray. Like Greater Yellowlegs and several other close relatives, frequently nods or teeters.

SIMILAR SPECIES. (1) Greater Yellowlegs is larger, with longer, usually up-turned bill. (2) Stilt Sandpiper has green or yellow legs with bill tip slightly decurved. (3) Solitary Sandpiper has a conspicuous eye ring, shorter bill, and boldly marked tail.

RANGE. Northern North America, wintering south to Argentina; West Indies. **Status**: Common winter resident and transient in Cuba, Isla de Pinos, and several cays. Found throughout the year. **Habitat**: Coastal lagoons, flooded fields, marshes. **Voice**: Quieter than Greater Yellowlegs, typically a flat *tu* or *tu-tu*. **Food**: Mostly terrestrial and aquatic insects, crustaceans.

SOLITARY SANDPIPER

ZARAPICO SOLITARIO

Tringa solitaria   Pls. 17, 19

DESCRIPTION. 8.5" (22 cm). Winter plumage mottled grayish brown above, with white belly, throat, and chest with brownish streaks. In breeding plumage, back heavily spotted with white; throat and chest heavily streaked with black. In all plumages, has a distinct white eye ring and tail boldly barred with black and white. Straight bill; olive green legs. In flight, wings unmarked above, strikingly dark below. Juvenile resembles adult, with less defined streaks on chest; upper chest has brownish wash. Wing-beats normally deep and jerky. Nods or teeters frequently.

SIMILAR SPECIES. (1) Spotted Sandpiper is white below and lacks grayish or white speckling above; white wing stripe is conspicuous during fluttering flight. (2) Stilt Sandpiper has longer bill, slightly decurved at tip, and lacks eye ring. (3) Lesser Yellowlegs lacks eye ring and has longer, yellow legs.

RANGE. Breeds in northern North America, wintering south to Argentina; West Indies. **Status**: Common winter resident and transient in Cuba, Isla de Pinos, and some cays (21 Jul–20 May). **Habitat**: Lagoons and ponds within forests, shores of rivers, lakes, estuaries. **Voice**: A rapid, high-pitched whistle, *peet-weet-weet*. **Food**: Aquatic and terrestrial insects, spiders, worms, mollusks.

WILLET
ZARAPICO REAL

Catoptrophorus semipalmatus Pls. 19, 20

DESCRIPTION. 15" (38 cm). Winter plumage pale gray above, with gray breast and white belly. Breeding plumage grayish brown, mottled with black above; breast and flanks barred and spotted. Bill long, rather thick; legs gray. In flight, shows a very striking white wing stripe. Juvenile similar to breeding adult but with whitish underparts.

SIMILAR SPECIES. (1) Hudsonian Godwit has very long, thin, and slightly upturned bill, less distinct wing pattern, and black tail. (2) Greater Yellowlegs has yellow legs and lacks white wing stripe.

RANGE. North America, wintering south to Chile and Brazil. In the Bahamas, Antilles (Cuba, Beata Island, Anegada, St. Croix, Antigua, Barbuda, St. Martin, and Anguilla). **Status**: Common permanent and winter resident; also a transient in Cuba, Isla de Pinos, and several cays. **Habitat**: Marshes, swamps, flooded fields, coastal lagoons, sandy beaches. **Nesting**: Apr–Jul. Nest is shallow scrape lined with vegetation or shells. Lays four pale-green eggs, finely speckled and spotted with dark brown, pale gray or purple. **Voice**: Noisy; in flight, *whee-wee-wee*; during the nesting season, *pill-will-willet*. **Food**: Aquatic insects, worms, small crabs, small mollusks, fish eggs, small fish.

SPOTTED SANDPIPER
ZARAPICO MANCHADO

Actitis macularia Pls. 19, 20

DESCRIPTION. 7.5" (19 cm). Brown above; small bill. Winter adult white below, with gray extending from neck onto breast, forming a white mark in front of wing; legs greenish. In breeding plumage, the underparts are marked with round dark-brown spots, more extensive on females than on males; legs flesh pink. In both plumages, rump is dark and wing stripe distinct. Juvenile resembles winter adult; wing coverts are thinly barred with black. Constantly raises and lowers the rear part of the body, not unlike Solitary Sandpiper but more emphatic and undulating. In flight, wingbeats are very stiff and shallow, interspersed with short glides. Usually solitary.

SIMILAR SPECIES. (1) Solitary Sandpiper has white-spotted back, heavily streaked chest, and unstriped wings with black wing linings. (2) Sanderling is pale gray above, with black legs and a small black patch on shoulder.

RANGE. North America, wintering south to Argentina; West Indies. **Status**: Common winter resident and transient in Cuba, Isla de Pinos, and some cays (9 Jul–23 May). **Habitat**: Rocky or muddy coasts, shores of lagoons, rivers. **Voice**: A measured *peet-weet-weet-weet-weet*, lower and slower than Solitary Sandpiper. **Food**: Worms, insects, crustaceans.

UPLAND SANDPIPER

GANGA

Bartramia longicauda Pl. 17

DESCRIPTION. 12" (30 cm). Mottled brown, with streaked throat and breast and white belly. Rather small head and bill; large eye; slender neck. Long tail and wings. Yellow legs. In flight, outer wing distinctly blackish; wingbeats shallow and stiff. Underwing coverts barred with black and white. Usually solitary. Juvenile resembles adult; scapulars blackish brown.

SIMILAR SPECIES. Buff-breasted Sandpiper is smaller, with shorter and thicker neck and shorter tail. In flight, underwing is white.

RANGE. North America, wintering in South America from Surinam and northern Brazil south to Argentina and Uruguay. **Status**: Very rare transient (Aug–Oct; Apr–May). **Habitat**: Savannas, pastures, agricultural fields. **Voice**: Flight call is a rapid, liquid *guip-iip-iip*. **Food**: Seeds, terrestrial invertebrates.

WHIMBREL

ZARAPICO PICO CIMITARRA CHICO

Numenius phaeopus Pl. 17

DESCRIPTION. 17" (43 cm). Grayish brown, somewhat mottled, with striped crown. Long, thin, and decurved bill; gray legs. In flight, shows grayish brown underwing. Juvenile has more conspicuous mottling on back and wings.

SIMILAR SPECIES. Long-billed Curlew is larger, buffy brown, with no stripes on head and longer bill. In flight, has bright cinnamon underwing.

RANGE. North American and Eurasian Arctic, wintering south to southern Argentina; West Indies. **Status**: Very rare winter resident and rare transient in Cuba (Sep–25 Apr). **Habitat**: Mudflats, beaches, lagoon shores, flooded areas, savannas. **Voice**: In flight, a tittering *pip-pip-pip-pip-pip*. **Food**: Crustaceans, mollusks, worms, seeds.

LONG-BILLED CURLEW
ZARAPICO PICO CIMITARRA GRANDE; ZARAPICO DE PICO LARGO
Numenius americanus   Pl. 17

DESCRIPTION. 23" (58 cm). Large, cinnamon brown above; buffy below. Extremely long, thin, decurved bill; bluish gray legs. In flight, shows a bright cinnamon underwing. Usually solitary. Juvenile has somewhat shorter bill.
SIMILAR SPECIES. (1) Whimbrel is grayish brown with striped crown, whitish belly, shorter bill, and grayish brown underwing. (2) Marbled Godwit has straight or slightly upturned bill.
RANGE. Western and Central North America, wintering south to Costa Rica. **Status**: Vagrant. Six records: Jun, Jul, Sep, Oct (last century); Sep (1967); 2 Sep (1987). Santa Fé, Cayo Romano. **Habitat**: Sandy beaches. **Voice**: Noisy; a piercing whistled *wee-it*. **Food**: Mostly insects, worms, crustaceans, toads.

HUDSONIAN GODWIT
AVOCETA PECHIRROJA
Limosa haemastica   Pl. 17

DESCRIPTION. 15" (38 cm). Winter adult gray brown above, with grayish brown wash on neck and upper breast and whitish belly. Bill very long, bicolored, and slightly upturned. Breeding plumage blackish and mottled above, reddish cinnamon below. In both plumages, black underwing is conspicuous in flight; also have a distinct white eyebrow, white uppertail coverts, and median stripe on wing, and black tail with broad white stripe at base and narrow white stripe at tip. Juvenile resembles breeding adults, but paler, with a scaly brownish black back and buffy underparts.
SIMILAR SPECIES. Willet has shorter, heavier, blunter, straight bill, and a very broad and bold white wing stripe.
RANGE. Breeds in northern North America, wintering in southern South America. **Status**: Vagrant. Three records: last century; 22 Sep; 5 Nov. Western Cuba. **Habitat**: Flooded fields, marshes, sandy shores. **Voice**: Usually silent in migration. Occasionally utters *kvick-kvick* or *quick-quick*. **Food**: Mostly marine invertebrates.

MARBLED GODWIT
AVOCETA PARDA
Limosa fedoa   Pl. 17

DESCRIPTION. 18" (46 cm). Large, buffy brown, mottled above, with a very long, straight, bicolored bill, often slightly upturned. In flight, shows cinnamon underwings. Winter adults pale brown below with little barring. In breeding plumage, underparts are heavily barred, reddish brown. Juvenile resembles winter adult, with a few bars on flanks.

SIMILAR SPECIES. Hudsonian Godwit is mostly grayish brown above, with whitish belly and black and white tail. In flight shows narrow white wing stripe.

RANGE. Central North America, wintering south along both coasts to Colombia, rarely to northern Chile. **Status**: Vagrant. Four records: Sep (last century); 12 Apr (1942, 1953); 20 Feb (1991). San Cristobal, Isabela de Sagua. **Habitat**: Savannas, lagoon shores, beaches, estuaries. **Voice**: A nasal *god-wit*. **Food**: Mostly mollusks, snails, crustaceans, worms, leeches.

RUDDY TURNSTONE

REVUELVEPIEDRAS

Arenaria interpres   Pls. 17, 19

DESCRIPTION. 9" (23 cm). Winter adult has brown upperparts and chest. Summer plumage is black, white, and reddish brown above. In flight, both plumages show a bold, complicated dorsal pattern of white markings on wings and rear body. Bill short, pointed, and slightly upturned; legs short, bright orange. Juvenile resembles winter adults, with wing coverts and scapulars edged with chestnut, and yellowish brown to orange legs. Flips small stones and shells with bill in search of food. Commonly flocks.

SIMILAR SPECIES. None.

RANGE. High latitudes of North America and Eurasia. American populations winter on coasts to southern South America; West Indies. **Status**: Common winter resident and transient in Cuba, Isla de Pinos, and many cays (3 Aug–20 May). **Habitat**: Rocky, sandy, and muddy seashores, particularly where marine debris accumulates. **Voice**: A guttural rattle, *kut-a-kut*. **Food**: Worms, mollusks, snails, crustaceans, carrion.

RED KNOT

ZARAPICO RARO

Calidris canutus   Pls. 18, 19

DESCRIPTION. 10.5" (27 cm). Stocky looking, with short bill, neck, and legs. Bill straight; legs dull green. Winter adults are gray above, whitish below, with chest finely streaked and washed with gray, and whitish supercilium. Breeding plumage dark brown above, rusty below. Juvenile resembles winter adult, with a scaly-looking back and a faint buffy wash below. In flight, both adult and juvenile show barred rump and a narrow wing stripe.

SIMILAR SPECIES. (1) Short-billed and (2) Long-billed Dowitchers have much longer bills and a white wedge extending from barred tail to lower back. (3) Sanderling in winter plumage is paler, lacks whitish supercilium, and shows black on shoulder. In breeding plumage, reddish coloration of underparts is suffused with black spots and restricted to breast.

RANGE. Breeds in the high Arctic, wintering to Chile and southern Argentina. **Status**: Rare and local winter resident and transient in Cuba (Aug; 15 Jan–26 Apr). **Habitat**: Brackish lagoons, coasts. **Voice**: A low *kuh*. **Food**: Mollusks, snails, marine worms, fish.

SANDERLING
ZARAPICO BLANCO
Calidris alba   Pl. 18

DESCRIPTION. 8" (20 cm). Winter adult strikingly pale, gray above, white below, with a small black patch on shoulder and black legs. In breeding plumage, upperparts and chest are reddish brown. Juvenile resembles winter adults, with black spots above. Bill straight, short, and black. In flight, shows a bold white wing stripe. Actively runs on wet sand following each receding wave, often in flocks.

SIMILAR SPECIES. (1) Winter Red Knot is larger, darker, with whitish supercilium and lacks black shoulder. In breeding plumage, has more extensive and unspotted red coloration below. (2) Winter Spotted Sandpiper is darker and browner above, with green legs and gray on breast highlighting a white area in front of wing.

RANGE. Breeds in northernmost North America and Eurasia. American populations winter south to Argentina; West Indies. **Status**: Common winter resident and transient in Cuba, Isla de Pinos, and several cays (7 Sep–Apr). **Habitat**: Sandy and sometimes muddy beaches. **Voice**: A sharp *plick*. **Food**: Marine invertebrates, spiders.

SEMIPALMATED SANDPIPER
ZARAPICO SEMIPALMEADO
Calidris pusilla   Pl. 18

DESCRIPTION. 6" (15 cm). Winter adult gray brown above, white below, with a straight, short, and rather thick bill, slightly club tipped. Breast lightly streaked. Legs black, toes partially webbed. Breeding adults are browner above and more heavily streaked on chest. Juvenile resembles breeding adult, but is more crisply marked above, but unstreaked below; distinct eyebrow.

SIMILAR SPECIES. (1) Least Sandpiper has longer and finer bill, slightly decurved at tip, and greenish to yellowish legs. Chest more conspicuously streaked. (2) Western Sandpiper is usually slightly paler above in winter plumage but has brighter rufous in juvenile and breeding plumage. Bill is usually longer, more decurved at tip, particularly in females. In winter plumage, only extremes are safely separated. (3) White-rumped Sandpiper has longer wings extending beyond tip of tail, and white rump.

RANGE. Breeds in northern North America, wintering south to northeastern South America; West Indies. **Status**: Common winter resident and

transient in Cuba, Isla de Pinos, and some cays (Aug–May). **Habitat**:
Mudflats along lagoons, estuaries, and seashores. **Voice**: A sharp *chit*, and
a murmuring and twittering while feeding. **Food**: Aquatic insects, worms,
crustaceans.

WESTERN SANDPIPER

ZARAPICO CHICO

Calidris mauri   Pl. 18

DESCRIPTION. 6.5" (17 cm). Winter adult has gray upperparts, whitish below, with finely streaked chest. Breeding plumage brown mottled above
with black, with chestnut on scapulars, ear patch, and crown; chest heavily
streaked with spotted sides. Bill long, decurved at tip; legs black and toes
partially webbed. Juvenile resembles breeding adult but with finely
streaked chest and buffy wash on upper chest.

SIMILAR SPECIES. (1) Least Sandpiper has heavily streaked chest and green
to greenish yellow legs. (2) Semipalmated Sandpiper has shorter and
straighter bill, particularly evident in males and is usually slightly darker
above in winter plumage. In juvenile and breeding plumage, appears
browner. In winter plumage, only extremes are safely separated. (3) White-
rumped Sandpiper has longer wings extending beyond tip of tail, and
white rump. (4) Dunlin is larger and darker above.

RANGE. Breeds in Alaska and northeastern Siberia. American populations
winter south to Peru; West Indies. **Status**: Rare or uncommon transient
and winter resident in Cuba, probably often overlooked (Jul–15 Apr).
Habitat: Mudflats along lagoons, estuaries, and seashores. **Voice**: A high
chit, and a murmuring and twittering while feeding. **Food**: Mostly marine
invertebrates.

LEAST SANDPIPER

ZARAPIQUITO

Calidris minutilla   Pls. 18, 19

DESCRIPTION. 6" (15 cm). Winter adult mottled grayish brown above, white
below; chest heavily streaked; legs green to greenish yellow. Bill tip slightly
drooped. Breeding plumage streaked below to upper belly, with darker back
and yellow legs. Juvenile resembles breeding adult, with scaly, chestnut-pat-
terned mantle; breast lightly streaked brown, with a buffy wash. Gathers in
large flocks, usually foraging higher on shore than other small shorebirds.

SIMILAR SPECIES. (1) Semipalmated Sandpiper has short, straight bill, black
legs, and less well-marked chest. (2) Western Sandpiper is grayer above,
with very pale chest and longer bill, often drooped at tip. (3) White-
rumped Sandpiper is larger and grayer, and has longer wings extending beyond tip of tail, and white rump.

RANGE. Breeds in northern North America, wintering south to northern South America; West Indies. **Status**: Common winter resident and transient in Cuba, Isla de Pinos and several cays (5 Jul–30 May). **Habitat**: Muddy shores of freshwater or salt water, flooded fields, beaches. **Voice**: A high-pitched, slightly trilled *breeeep*. **Food**: Crustaceans, mollusks, worms, seeds.

WHITE-RUMPED SANDPIPER

ZARAPICO DE RABADILLA BLANCA

Calidris fuscicollis   Pls. 18, 19

DESCRIPTION. 7″ (18 cm). Winter adult mottled brownish gray above, white below, with blackish bill and legs, streaked chest and sides, and fairly pronounced white eyebrow. White rump and faint wing stripe visible in flight. Breeding plumage blackish brown above, with more distinct black streaks on chest and sides. Juvenile resembles breeding adult, with chest and sides washed buffy gray and less distinctly streaked. When standing, wing tips extend well beyond tail, giving a distinctly long, tapered silhouette to the rear body.

SIMILAR SPECIES. (1) Semipalmated and (2) Western Sandpipers are paler and have shorter wings not projecting beyond tip of tail, and dark rumps. (3) Least Sandpiper is smaller, browner, with green to yellowish green legs and dark rump. (4) Dunlin has slightly streaked breast, bill decurved at tip, and dark rump.

RANGE. Breeds in northern North America, wintering in South America south to Argentina. **Status**: Rare transient in Cuba (Jul–Nov; 15 Apr–5 Jun). **Habitat**: Muddy fields, rice fields, beaches. **Voice**: In flight, *jeet*, very high pitched. **Food**: Seeds, worms, snails.

PECTORAL SANDPIPER

ZARAPICO MOTEADO

Calidris melanotos   Pl. 18

DESCRIPTION. 9″ (23 cm). Winter adult mottled grayish brown above, white below; the border between the dark chest and white belly is sharp. Bill rather short, slightly decurved; legs yellowish green. In breeding plumage, the back is darker, the chest is heavily streaked, and legs are yellow. Juvenile similar to breeding adult, with breast strongly washed buff.

SIMILAR SPECIES. Buff-breasted Sandpiper is more warmly colored with buff all over and lacks the pronounced breast band streaking.

RANGE. Breeds in Siberia and northwestern North America. American populations winter in South America from Peru to Argentina. **Status**: Rare transient in Cuba and some cays (Jul–Dec; 12 Mar–10 May). **Habitat**: Wet grassy areas, savannas, pastures, shores of lagoons. **Voice**: Noisy, especially when taking flight in compact flocks; a rich, reedy *brrrrrp*. **Food**: Spiders, worms, seeds.

DUNLIN

ZARAPICO GRIS

Calidris alpina   Pls. 18, 19

DESCRIPTION. 8.5" (22 cm). Winter adult brownish gray above and white below, with lightly streaked breast washed with gray; eyebrow white. Neck short; bill decurved at tip; legs black. Breeding birds have rusty upperparts, and whitish underparts with finely streaked breast and distinct black patch on belly. Juvenile has conspicuously streaked and spotted underparts. All plumages have dark rump.

SIMILAR SPECIES. (1) Semipalmated and (2) Western Sandpipers are smaller and mostly white below and have shorter bills. (3) Sanderling has straight bill, white underparts, and black patch on shoulder. (4) White-rumped Sandpiper has distinctly streaked chest and sides, white rump, and longer wings.

RANGE. Breeds in high Arctic North America and Eurasia. American populations winter south to Mexico. **Status**: Very rare transient and winter resident at Zapata peninsula and Cayo Coco. Six records: 2 Jan–21 Mar. **Habitat**: Estuaries, lagoons, beaches. **Voice**: A rough *jeeeep*. **Food**: Spiders, worms, mollusks.

STILT SANDPIPER

ZARAPICO PATILARGO

Calidris himantopus   Pls. 18, 19

DESCRIPTION. 8.5" (22 cm). Winter adult brownish gray above and white below, with lightly streaked neck and chest; well-defined white supercilium. Bill long, slightly decurved, and thickened at tip; legs yellowish green. Breeding plumage is blackish above, with a chestnut cheek patch; chest and belly barred with dark brown. Juvenile resembles breeding adult above but is unbarred below and lacks distinct chestnut cheek patch. In flight, wing is unmarked and rump is white. While feeding, plunges bill vertically into sediment with rapid "sewing machine" motion, similar to the dowitchers, which it often accompanies.

SIMILAR SPECIES. (1) Short-billed and (2) Long-billed Dowitchers are larger with longer straight bills; wedge-shaped patch of white extends from rump onto back. (3) Lesser Yellowlegs is larger, with a straight bill and bright yellow legs. (4) Solitary Sandpiper has shorter straight bill, distinct eye ring, boldly barred tail, and striking black underwing.

RANGE. Breeds in northern North America, wintering in South America from Bolivia to Argentina; West Indies. **Status**: Common transient in Cuba (Jul–Dec; 11 Feb–5 Jun). **Habitat**: Shallow pools, salt water or freshwater. **Voice**: A hoarse *querp*. **Food**: Worms, mollusks, seeds, roots.

BUFF-BREASTED SANDPIPER

ZARAPICO PIQUICORTO

Tryngites subruficollis   Pl. 20

DESCRIPTION. 8.5" (22 cm). Mottled brown above; uniformly buff below. Underwing silvery white. Bill short and brownish black, paler at base; long, yellow legs. Juvenile is paler.

SIMILAR SPECIES. (1) Pectoral Sandpiper is grayer above, less buffy below, and has streaked breast band contrasting strongly with white belly. (2) Upland Sandpiper is larger, with longer neck and tail, larger eye, and barred underwing.

RANGE. Breeds in northwestern North America, wintering south to Argentina. **Status**: Vagrant. One record: last century (Gundlach's collection). **Habitat**: Low vegetation savannas, cultivated fields. **Voice**: A low trilling *preeet* when flushed. Seldom vocalizes on migration or on wintering grounds. **Food**: Spiders, aquatic vegetation, seeds.

SHORT-BILLED DOWITCHER

ZARAPICO BECASINA

Limnodromus griseus   Pls. 19, 20

DESCRIPTION. 11" (28 cm). Winter adults are brownish gray above, paler on neck, with white eyebrow and belly. Chest gray, finely streaked. Bill long, straight, and thick. Breeding plumage has upperparts mottled with dark brown; neck and sides reddish brown. Juvenile resembles breeding adults, but brownish; indistinctly marked below; tertials with rusty pattern to feather centers. In flight, a sharp spike of white stretches from uppertail to middle of back. Probes sediment with "sewing machine" motion.

SIMILAR SPECIES. (1) Winter plumage Long-billed Dowitcher is best distinguished by call. Bill of Long-billed Dowitcher is usually slightly longer; throat to lower breast is dark gray and unstreaked; has different habitat preferences. Juvenile has dark tertials with rusty edging. (2) Common Snipe is dark brown and white; head and back boldly striped. (3) Stilt Sandpiper has shorter, thinner, slightly decurved bill; white is restricted to rump.

RANGE. Breeds in northern North America, wintering south to Peru and Brazil; West Indies. **Status**: Common winter resident and transient in Cuba, Isla de Pinos, and some cays (Jun–Apr). **Habitat**: Mainly near seashores, exceptionally in lagoons and flooded fields. **Voice**: When flying, a rapid whistled *tu-tu-tu*. **Food**: Aquatic insects, mollusks, crustaceans, marine worms, seeds.

LONG-BILLED DOWITCHER

ZARAPICO BECASINA DE PICO LARGO

Limnodromus scolopaceus   Pl. 20

DESCRIPTION. 11.5" (29 cm). Very similar to Short-billed Dowitcher, whose bill may be as large as that of this species. The two are best separated by voice. Breeding plumage has uniformly brownish-red underparts; heavily spotted breast, and barred sides. Juvenile resembles winter adult, but darker on back.

SIMILAR SPECIES. (1) Winter plumage Short-billed Dowitcher is best distinguished by call. Short-billed Dowitcher has paler, finely streaked breast and different habitat preferences. Juvenile has tertials with rusty pattern in center of feathers. (2) Common Snipe is dark brown and white; head and back are boldly striped. (3) Stilt Sandpiper has shorter, thinner, slightly decurved bill; white is restricted to rump.

RANGE. Breeds in extreme northwestern North America and northeastern Siberia. American populations winter south to Panama. **Status**: Rare transient and very rare winter resident in Cuba (Jan–18 Apr). **Habitat**: Shores of lagoons and swamps, flooded savannas. **Voice**: A single, high-pitched *keek*, sometimes uttered in a rapid series. **Food**: Aquatic insects, mollusks, crustaceans, marine worms, spiders.

COMMON SNIPE

BECASINA

Gallinago gallinago   Pls. 19, 20

DESCRIPTION. 11" (28 cm). Dark brown above, with striped head and back, white belly, broadly streaked breast, and heavily barred sides. Wings long and pointed; tail rusty. Bill very long, straight, and rather thick. Camouflage patterning makes it difficult to spot on the ground, among grasses. Flies with very rapid wingbeats, frequently zigzagging when arising, and uttering its distinctive call. Not normally found in flocks.

SIMILAR SPECIES. (1) Short-billed and (2) Long-billed Dowitchers have unstriped heads and a distinct white patch from rump to middle of back.

RANGE. Eurasia, North and South America. North American populations winter to southern South America and West Indies. **Status**: Common winter resident in Cuba, Isla de Pinos, and Cayo Coco (19 Aug–May). **Habitat**: Swamps, lagoons, flooded savannas, brackish shores. **Voice**: A harsh *rehk* when flushed. **Food**: Insects, mollusks, crustaceans.

WILSON'S PHALAROPE*

ZARAPICO DE WILSON

Phalaropus tricolor   Pl. 17

DESCRIPTION. 9" (23 cm). Winter adult gray above with white uppertail coverts; white below. Crown gray; lores and eye stripe indistinctly gray. Cheeks practically unmarked; wings entirely gray. Bill long and notably thin; legs yellow. In breeding plumage, legs are black; female has rich cinnamon stripe along neck and back. Male is duller. Juvenile resembles winter plumage adults, but browner above, with buffy breast. Often swims but seen on land much more frequently than other phalaropes. More common inland.

SIMILAR SPECIES. Both (1) Red-necked and (2) Red Phalaropes have shorter bills, black auricular patches, and wing stripes. Both are mainly marine in nonbreeding season.

RANGE. Central and western North America, wintering in South America from Peru and Bolivia south to Argentina. **Status**: Two unconfirmed records: Salina de Bidos, Matanzas province; and Gibara, Holguín province. **Habitat**: Shallow, freshwater or brackish pools. **Voice**: A high-pitched *creep*. **Food**: Aquatic insects, crustaceans, spiders, seeds.

RED-NECKED PHALAROPE

ZARAPICO NADADOR

Phalaropus lobatus Pl. 17

DESCRIPTION. 7" (18 cm). Winter adult gray above with blackish nape and whitish stripes on back, white below, with a black patch through eye, and a white stripe on wing. Bill thin. In breeding plumage, both sexes have reddish brown stripes on back and patches on sides of neck. Male is duller. Juvenile resembles breeding adults, but duller. Virtually always seen swimming.

SIMILAR SPECIES. (1) Red Phalarope has uniformly gray back and thicker bill. (2) Wilson's Phalarope has longer bill and lacks wing stripe; is more likely to be found in freshwater.

RANGE. Breeds in northern North America, Greenland, and northern Eurasia. American populations winter in South America. **Status**: Vagrant. Two records: 20 May (1953); 10 Dec (1963). Gibara and Cienfuegos. **Habitat**: Brackish pools, estuaries, occasionally at sea. **Voice**: A soft *krit* or *pick*. **Food**: Aquatic insects, crustaceans, mollusks, zooplankton.

RED PHALAROPE

ZARAPICO ROJO

Phalaropus fulicaria Pl. 17

DESCRIPTION. 8.5" (22 cm). Winter adult pale gray above with black nape, white below, with black patch through eye, and a white stripe on wing. Bill thick. In breeding plumage, cheeks white; underparts uniformly reddish brown. Male is paler. Juvenile similar to breeding male, but paler. Virtually always seen swimming.

SIMILAR SPECIES. (1) Red-necked Phalarope is darker above, with striped back, and thinner bill. (2) Wilson's Phalarope has longer bill and practically unmarked cheek and auricular. In flight, wings lacks wing stripe. More likely to be found in freshwater.

RANGE. Breeds in northern North America, in the Palearctic from Greenland and Iceland east through Arctic islands to northern Siberia. North American populations winter south to South America. **Status**: Vagrant. Two records: 10 Dec (1963); 30 Jan (1967). Gibara. **Habitat**: Essentially pelagic outside breeding season; occasionally estuaries and inland lakes. **Voice**: A sharp *peek*. **Food**: Mostly aquatic insects, zooplankton.

JAEGERS, SKUAS, GULLS, TERNS, AND SKIMMERS
Laridae

Jaegers and gulls are small to large, rather heavy-bodied birds with slightly hooked bills; terns are lighter-bodied, with narrow, sharp-pointed wings, and long straight bills. The feet are webbed. Most species are pale gray and white; some have dark gray to black backs and others are entirely dark brown. Several juvenile and first-winter gull and tern species have a brown bar on the wing; young gulls have a black terminal tail band, and young terns lack elongated outer tail feathers ("streamers"). Diet is mainly fish. Jaegers often obtain food by harassing other seabirds; gulls pick from the surface while fluttering, or feed on shores; terns hover over the water's surface, making shallow headfirst plunges, and spend more time on the wing. Jaegers and gulls also scavenge and consume carrion. In flight, the bill is usually held horizontally or in terns pointed slightly downward. Breeding plumage in some species includes the development of a hood or crown, and color changes in the feet and bill. Jaegers occur in light, dark, or intermediate color morphs. Sexes alike. (**W**:129; **C**:27)

SOUTH POLAR SKUA
SKUA DEL POLO SUR

Catharacta maccormicki   Pl. 20

DESCRIPTION. 23" (58 cm). Dark brown, except for very bold white patch at base of primaries; densely streaked upperparts and underparts. Wings very broad. Immature has less streaked underparts. Flight powerful, and habits, like jaegers, piratical, pursuing other seabirds.
SIMILAR SPECIES. (1) Jaegers are smaller, with narrower wings and longer tails. (2) Juvenile Herring Gull is paler and lacks white wing patches.
RANGE. Breeds on the South Shetland Islands and Antarctica; ranges at sea regularly to the North Pacific and to the North Atlantic. **Status**: Vagrant. One record: 13 Dec (1986). Bahía de Nuevitas. **Habitat**: Oceanic. **Voice**: Usually silent in winter. **Food**: Fish.

POMARINE JAEGER
ESTERCORARIO POMARINO

Stercorarius pomarinus   Pl. 20

DESCRIPTION. 21" (53 cm). Mostly very dark brown with broad wings. Two central tail feathers elongated, twisted, and spoon shaped. Like other jaegers, light-morph birds are whitish below with a dark band across the breast, somewhat heavier in this species than in Parasitic Jaeger; a black cap with a wash of yellow on nape and sides of neck; and dark barring along sides and flanks. Dark-morph birds are dark brown below. In both morphs, there is a conspicuous white patch at the base of the primaries, more extensive below than above. On the underwing, an additional pale patch at the base of the primary coverts is diagnostic. Winter adults resemble subadults and have barring on rump and undertail coverts, narrow light barring on back, and in light-phase birds, more extensive and uneven markings on underparts; black cap is less sharply defined and central tail feathers are shorter. Juvenile has square-ended and short central tail feathers; light-morph individuals are neatly barred below. Flight gulllike, somewhat more leisurely than Parasitic Jaeger. Usually detected when engaged in spectacular twisting and diving pursuit of other seabirds, such as gulls and terns.

SIMILAR SPECIES. (1) Parasitic Jaeger is smaller, with pointed and rather short tail feathers. Wings appear narrower at base, with white at base of primaries less conspicuous and pale area at base of underwing primary coverts lacking. Juvenile less neatly marked below. (2) Long-tailed Jaeger has very long and pointed central tail feathers, and an all-white breast. White on upperwing is inconspicuous or lacking. Juvenile has conspicuously barred undertail and underwing coverts. (3) Great Skua is larger, with broader wings, and shorter tail.

RANGE. Breeds in Arctic latitudes of North America and Eurasia, wintering south to Southern Hemisphere seas; West Indies. **Status**: Very rare winter visitor to Cuban seas (1 Nov–22 Apr). **Habitat**: Pelagic. **Voice**: Usually silent in winter. **Food**: Fish, offal, carrion.

PARASITIC JAEGER

ESTERCORARIO PARASÍTICO

Stercorarius parasiticus Pl. 20

DESCRIPTION. 17″ (43 cm). Very similar to Pomarine Jaeger. Central tail feathers sharply pointed. Light-morph birds have dark cap, white forehead, white chest and upper belly, and gray lower belly. A pale breast band may be present. Upperwing with white patch as in Pomarine Jaeger, but somewhat less conspicuous. Dark-morph birds entirely dark brown below. Winter adult resembles subadults and has barred rump and undertail coverts, narrow light barring on back, and in light-phase birds, more extensive and uneven markings on underparts; black cap is less sharply defined and central tail feathers are shorter. Juvenile barred dusky below, more heavily on undertail coverts; central tail feathers protrude visibly. Flight is fast, falcon like. Like Pomarine Jaeger, it is usually spotted when pursuing other seabirds.

SIMILAR SPECIES. (1) Pomarine Jaeger is larger, the wings appearing significantly broader at base; projecting central tail feathers broad and blunt; breast band is broad and dark. Pale area at base of underwing primary coverts is diagnostic. Juvenile strongly barred below, with a more distinct white patch above primaries. (2) Long-tailed Jaeger has narrow wings, very long and pointed central tail feathers, and an all-white breast. White on upper primaries practically absent. Juveniles of both color morphs have conspicuously barred underwing and undertail coverts. (3) Great Skua is larger, with broader wings and shorter tail.

RANGE. Breeds in Arctic latitudes of North America and Eurasia, wintering to Southern Hemisphere oceans. **Status**: Very rare winter visitor to Cuban seas (Jul–28 Oct; Mar–14 Apr). **Habitat**: Oceanic. **Voice**: Usually silent in winter. **Food**: Mostly fish, largely by piracy.

LONG-TAILED JAEGER
ESTERCORARIO RABERO
Stercorarius longicaudus   Pl. 20

DESCRIPTION. 16" (41 cm). Gray above, with contrasting darker flight feathers. Light-morph has white chest, darkening to gray toward tail; a black cap; and a yellow wash on sides of neck and nape. Rare dark morph is entirely dark brown below. Central tail feathers very long and sharply pointed. Winter adult has distinct barring on undertail coverts and flanks, like juvenile; gray collar and indistinct crown patch; and shorter central tail feathers. Juveniles of both color morphs have short central tail feathers, strong barring on underwing and undertail coverts, dark upperwing, and a whitish patch on underwing. Slender, with airy, ternlike flight.

SIMILAR SPECIES. (1) Pomarine Jaeger has short and broad central tail feathers, broader wings with a white patch above, and a dark breast band. Juvenile darker above with stronger barring below and a distinct white patch on primaries. (2) Parasitic Jaeger has sharp, but short central tail feathers, broader wings with a white patch above, and usually a brownish breast band. Juvenile has white patch on outer wing. (3) Great Skua is larger, with broader wings and shorter tail.

RANGE. Breeds in Arctic latitudes of North America and Eurasia, wintering south to Southern Hemisphere oceans. **Status**: Vagrant. Two records: 29 Nov (1937); Mar (1953). Caibarién and Gibara. **Habitat**: Oceanic. **Voice**: Usually silent in winter. **Food**: Birds, pirated fish, offal.

LAUGHING GULL
GALLEGUITO
Larus atricilla   Pl. 21

DESCRIPTION. 16" (41 cm). White with dark gray mantle and black wing tips. In breeding plumage, head black with white eye ring; legs and rather long bill dark red. In winter plumage, head mostly white, with wash of gray around and behind eye; legs and bill black. Juvenile has dull brown upperparts and white rump.

SIMILAR SPECIES. (1) Black-headed Gull is smaller, with much paler gray mantle and distinct white patch on leading edge of wing. (2) Bonaparte's Gull differs in the same ways, and is even smaller.

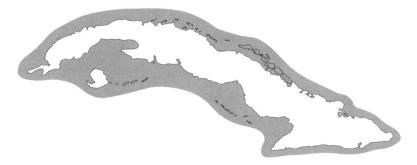

RANGE. Atlantic coasts from Nova Scotia to French Guiana; also southern California to Mexico; West Indies. **Status**: Permanent resident and the most common gull along Cuban coasts. Also a winter resident and transient. **Habitat**: Waters around cays, bays, beaches. **Nesting**: May–Jun. Builds crude nest of seaweed, feathers, shells, and other debris on sand or on bare rock, laying two to four olivaceous eggs, spotted with brown. **Voice**: Noisy; a loud, laughter-like *ha-ha-ha-haah-haah-haah*. **Food**: Garbage, snails, insects, seabird eggs.

BLACK-HEADED GULL

GALLEGUITO RARO

Larus ridibundus   Pl. 21

DESCRIPTION. 16″ (41 cm). Winter adult has pale gray mantle, a dark ear spot, white tail, and red bill and legs. In flight, shows distinct white wedge on leading edge of outer wing; undersurface of outer wing is dark smoke-gray. In breeding plumage, has brown hood and white eye ring. First-winter birds and juveniles have pale dark-tipped bill and pale legs.

SIMILAR SPECIES. (1) Bonaparte's Gull in winter has black bill, pale pink legs, and mostly white undersurface of outer wing. (2) Laughing Gull has dark gray mantle and black wing tips.

RANGE. Breeds throughout northern Europe and Asia, wintering in the Americas along Atlantic coast of North America from Labrador to New York, sparingly as far as northern South America. **Status**: Vagrant. Two records: dates unknown. Archipiélago de los Canarreos and Holguín. **Habitat**: Harbors, estuaries. **Voice**: Usually silent in winter. **Food**: Insects, worms, scavenges garbage.

BONAPARTE'S GULL

GALLEGUITO CHICO

Larus philadelphia   Pl. 21

DESCRIPTION. 13″ (33 cm). Winter adults are white; mantle pale gray, with a white wedge on outer wing and mostly white underwing tips. Bill and ear spot black. In breeding plumage, head black and legs orange red. Juvenile has extensive black on wing tips.

SIMILAR SPECIES. (1) Laughing Gull has dark gray mantle. (2) Sabine's Gull has bold tricolored mantle of white, gray, and black, and forked tail. (3) Black-headed Gull is larger with red bill and darker undersurface of outer wing.

RANGE. Breeds in Canada and Alaska, wintering south to Mexico; Bahamas and Greater Antilles. **Status**: Very rare and local winter resident to Cuba (17 Aug–20 Mar). **Habitat**: Coasts. **Voice**: A harsh, somewhat nasal *tiar*. **Food**: Fish, crustaceans, worms.

RING-BILLED GULL

GALLEGO REAL

Larus delawarensis   Pl. 21

DESCRIPTION. 19" (48 cm). Somewhat larger than Laughing Gull. Winter adult white with pale gray mantle and white-spotted black wing tips. Nape and hindneck spotted with brown. Bill yellow with black ring near tip; legs yellowish green. Breeding adult similar, with head entirely white. In juvenile, the bill is black, later pink with dark tip, and the legs are pink.

SIMILAR SPECIES. (1) Juvenile Laughing Gull is much browner, with black bill and legs. (2) Herring Gull is considerably larger; subadult has broader subterminal tail band. (3) Black-legged Kittiwake has all-yellow bill, black legs, and completely black wing tips.

RANGE. Northern United States and Canada, wintering south to Mexico and rarely to northern South America; Bahamas and Greater Antilles. **Status**: Uncommon winter resident and transient in Cuban waters (13 Oct–13 May). **Habitat**: Coastal estuaries, lakes. **Voice**: A high-pitched *kree*, also a high yelping *yow*. **Food**: Fish, garbage, offal.

HERRING GULL

GALLEGO

Larus argentatus   Pl. 21

DESCRIPTION. 24" (61 cm). Winter adult white, streaked with brown on head and neck, with pale gray mantle; wing tips black, spotted with white. Bill yellow; legs flesh colored. Breeding adult has unmarked head and chest. Juvenile uniformly brown, with very dark brown tail and black bill. Older birds in winter are paler, with black-tipped bill and a broad subterminal tail band.

SIMILAR SPECIES. (1) First-winter Great Black-backed Gull is larger, with paler head and body and checkered mantle. (2) Ring-billed Gull is smaller; first-winter juvenile has mostly white tail, with only narrow dark band, and bicolored bill. (3) Great Skua vaguely resembles juvenile, but is darker, with white wing patches.

RANGE. Breeds in North America and Eurasia, wintering south to Panama; West Indies. **Status**: Uncommon winter resident and transient in Cuba, Isla de Pinos, and several cays (15 Sep–21 May). **Habitat**: Open water, bays, beaches. **Voice**: A loud *kyow*. **Food**: Scavenger; omnivorous.

GREAT BLACK-BACKED GULL

GALLEGÓN

Larus marinus   Pl. 21

DESCRIPTION. 30" (76 cm). White, with black mantle and wings. Bill heavy, yellow in breeding plumage, black-tipped in winter; legs flesh colored. Juvenile has mantle checkered in white and blackish brown, white rump, lightly streaked underparts, and barred brown undertail coverts. First-winter birds and juveniles have all-dark bill. In flight, broad wings have distinct white trailing edges and white tips.

SIMILAR SPECIES. First-winter Herring Gull is smaller and darker all over.

RANGE. Breeds on Atlantic coasts of North America and Europe. In the Americas, winters south to Florida. **Status**: Very rare winter visitor to Cuba, where most birds are subadults (Dec–Mar). Two individuals collected; six observed. **Habitat**: Inshore waters but also at sea. **Voice**: A hollow *cowwp*. **Food**: Fish, squid, offal.

BLACK-LEGGED KITTIWAKE

GALLEGO PATINEGRO

Rissa tridactyla   Pl. 21

DESCRIPTION. 17" (43 cm). Winter plumage white with gray wash on nape, light gray mantle, and black wing tips containing no white spots. A distinct black ear spot may extend to hindneck. Tail slightly forked. Bill yellow; legs black. Breeding plumage resembles winter adult, with head completely white. First-winter plumage has dark bars on lower hindneck, wings, and tail tip; black bill; and dark ear spot. Flight fast, with rapid wingbeats.

SIMILAR SPECIES. (1) Breeding Ring-billed Gull is larger, with fan-shaped tail and black ring on bill; black patch on wing tip contains two white spots. (2) Juvenile Sabine's Gull has an entirely black bill. Back and most of the inner wing are dark brown; wing tips are black and remainder of wings white, giving a tricolored appearance.

RANGE. Breeds in northern North America and Eurasia, wintering south to Baja California and Atlantic Florida. **Status**: Vagrant. Four records: 4 Jan (1949); Dec (1954); (1987); 12 Dec (1993). Offshore La Habana and Gibara. **Habitat**: Oceanic. **Voice**: A nasal *kitti-wake*, almost never heard away from colonies. **Food**: Aquatic invertebrates, offal.

SABINE'S GULL

GALLEGUITO DE COLA AHORQUILLADA

Xema sabini   Pl. 21

DESCRIPTION. 13.5" (34 cm). White. Upperwing divided into boldly contrasting triangles of white, gray, and black. Bill black with yellow tip; legs black. Tail rather deeply forked. In breeding plumage, head gray. Juvenile has entirely black bill. Back and most of innerwing dark brown; wing tips black and remainder of wing white, giving a tricolored appearance. A broad black band on tail tip.

SIMILAR SPECIES. (1) Bonaparte's Gull has pale gray mantle, lacking bold contrasts; fan-shaped tail; and pink or orange red legs. (2) First-winter Black-legged Kittiwake has dark bars on lower hindneck, wings, and tail tip.

RANGE. Breeds in northern North America, northern Greenland, Spitsbergen, and northern Siberia, wintering south to Chile. **Status**: Vagrant. One record: Dec (1954). Gibara. **Habitat**: Oceanic. **Voice**: Silent out of the breeding season. **Food**: Fish, mollusks.

GULL-BILLED TERN

GAVIOTA DE PICO CORTO

Sterna nilotica   Pl. 22

DESCRIPTION. 14" (36 cm). White with uniform, very pale gray mantle and rump, thick black bill, and rather long black legs. Wing somewhat broader than those of most other terns, but similarly pointed; tail only slightly forked. Black cap of breeding plumage is lost in winter, when head is mostly white with dusky auricular patch, and nape streaks may form a band from eye to eye. Juvenile has pale brown wash from crown to hindneck and a distinct dark mark on side of face.

SIMILAR SPECIES. (1) All other medium-sized, black-capped terns have longer, slimmer bills and deeply forked tails. (2) Forster's Tern has white rump and pink or orange legs, thinner bill, and a distinct silvery-white triangular patch on hindwing.

RANGE. Breeds locally in United States, the Caribbean, South America, Europe, Africa, and Asia. North American and West Indian birds winter at scattered locations in the West Indies and south along both coasts of Mexico and Central America to northern South America. **Status**: Rare transient and winter resident (Aug–14 Jun). **Habitat**: Lakes, lagoons, brackish ponds, flooded or plowed fields, occasionally inshore waters. **Voice**: A rasping, repeated *za* notes and *kay-weck*. **Food**: Crustaceans, insects, frogs.

CASPIAN TERN

GAVIOTA REAL GRANDE

Sterna caspia   Pl. 22

DESCRIPTION. 21" (53 cm). A large, white tern with pale gray mantle and heavy red bill, usually with a black tip, and black legs. Wing tips black below; tail only slightly forked. In breeding plumage, has black cap; winter adult and juvenile have streaked head and black nape. Juvenile has dull reddish orange bill and brown-tipped back feathers.

SIMILAR SPECIES. Royal Tern is slightly smaller, with somewhat more deeply forked tail and distinctly thinner orange bill; wings are paler below and slightly darker above. Forehead of Caspian Tern is never pure unstreaked white as in winter Royal Tern.

RANGE. Breeds locally in North and Central America, Eurasia, and Africa. In the Americas, winters south very sparingly to northern South America. Bahamas and Greater Antilles. **Status**: Uncommon winter resident on Cuban coasts. Apparently permanent resident at Zapata peninsula where locally common and possibly breeding (10 Oct–3 Jun). **Habitat**: Coastal waters, brackish lagoons, estuaries. **Voice**: A harsh *kraaow*. **Food**: Fish, crustaceans.

ROYAL TERN
GAVIOTA REAL
Sterna maxima   Pl. 22

DESCRIPTION. 20″ (51 cm). Large, white, with pale gray mantle, bright orange bill, and black legs. Wing tips black above; tail moderately forked. In breeding plumage, black cap forms bushy crest on nape; nonbreeding adult has white forehead and black nape. Juvenile resembles winter adult but is spotted above, with yellow to pale orange bill and legs.

SIMILAR SPECIES. Caspian Tern is slightly larger, with heavier red bill and dark underwing tips. Tail is less deeply forked. Hindcrown of Caspian Tern gives a squared-off impression, and winter adult has streaked forehead.

RANGE. Breeds locally on Atlantic and Pacific coasts of United States and Mexico, Atlantic coast of South America; West Indies; West Africa. North American populations winter south to Argentina and Cuba. **Status**: Common permanent resident, winter resident, and transient. **Habitat**: Coasts, beaches, bays. **Nesting**: May–Jul. Nests in sandy depression, in mixed colonies with Sandwich Tern. Lays single creamy-white egg, with spots and specks in black and gray. **Voice**: A loud rolling *kirrip* frequently repeated in flight. **Food**: Fish, crabs, squid, shrimp.

SANDWICH TERN
GAVIOTA DE SANDWICH
Sterna sandvicensis   Pl. 22

DESCRIPTION. 15″ (38 cm). White, with light gray mantle, moderately forked tail, and white rump. Bill black with yellow tip (visible only at close

range); legs black. In breeding plumage, cap and bushy nape black. Non-breeding adult and juvenile have white forehead and streaked crown. Juvenile has less deeply forked tail; bill may be entirely black; back has indistinct blackish bars.

SIMILAR SPECIES. Gull-billed Tern has short, heavy, and entirely black bill; less deeply forked tail; and broader wings.

RANGE. Breeds in eastern United States and Europe. In the Americas, winters south to northern Argentina; locally on some islands of West Indies. **Status**: Common summer resident and transient in Cuba, Isla de Pinos, and some cays. Local winter resident in northern cays. **Habitat**: Coasts, bays, beaches. **Nesting**: May–Jul. Nests in crowded colonies, with Royal Terns on small isolated cays. In Cuba, lays a single creamy-white egg, marked with spots, blotches, specks, and scrawls of brown and dark olive-brown, in a sandy depression. **Voice**: A loud, harsh *kurrick*. **Food**: Fish, shrimp, marine worms, squid.

ROSEATE TERN
GAVIOTA ROSADA
Sterna dougallii   Pl. 22

DESCRIPTION. 15.5" (39 cm). Very elongated, slender, and strikingly white. Mantle uniformly very pale gray; tail entirely white and deeply forked; underwing pure white, quite translucent. Wings short. Flies with rapid, shallow wingbeats. Breeding adult has black cap and nape; with a slight wash

of pink on throat and belly; black bill, red at base; and dark red legs. Non-breeding adult has black bill, brownish orange legs, and white forehead and crown. Juvenile has brownish cap, extending over forehead; scaly-looking mantle; and black bill and legs.

SIMILAR SPECIES. (1) Forster's Tern has darker underwings and a bold black stripe behind eye in winter plumage. (2) Gull-billed Tern has only slightly forked tail and much heavier bill. (3) Arctic Tern is darker overall, with black trailing edge on outer wing. Breeding adult has entirely red bill. Juvenile has black bar on upper front edge of wing. (4) Common Tern has less forked tail, with blackish sides, and darker wing tips. Juvenile has blackish bar on upper front edge of wing.

RANGE. Breeds locally in all major oceans, wintering widely across them; West Indies. **Status**: Rare summer resident and transient in Cuba. Breeds at Cayo La Vela near Sagua la Grande, Cayo Mono Grande near Varadero, and at Cayos Los Ballenatos near Cayo Largo. **Nesting**: May–Jul. Nests in small colonies. Lays a single cream egg, spotted with blackish brown, on bare rock. **Habitat**: Open water and bays near nesting cays. **Voice**: A soft and very distinctive *chivy* or *chuick*. **Food**: Fish, mollusks.

COMMON TERN
GAVIOTA COMÚN
Sterna hirundo   Pl. 22

DESCRIPTION. 14.5″ (37 cm). White, with pale gray mantle. Wing tips shading to dark gray; underwings light gray, with a broad blackish trailing edge on outer wing; tail deeply forked with blackish sides. Breeders have black cap and red bill with black tip. Legs and feet red. In nonbreeding plumage, forehead white; crown and nape black. Juvenile brownish above, white below, with mostly dark bill.

SIMILAR SPECIES. (1) In breeding plumage, Arctic Tern is grayer below; undersurface of outer wing has a narrow black border on trailing edge; tail is deeply forked; bill is entirely red. (2) Forster's Tern has pale underwing and a distinct silver-white triangular patch on rear of wing. Nonbreeder has thick black stripe behind eye. (3) Roseate Tern is distinctly whiter overall, and has very deeply forked tail.

RANGE. Breeds in North America, Eurasia, and West Indies. In the Americas, winters south to northern Argentina. **Status**: Rare transient (17 Aug–7 Jan). **Habitat**: Bays, coasts, lagoons. **Voice**: A repeated *kip*, and a harsh *kee-aa-rr*. **Food**: Fish, crustaceans.

ARCTIC TERN
GAVIOTA ÁRTICA
Sterna paradisaea   Pl. 22

DESCRIPTION. 15.5″ (39 cm). Gray below in breeding plumage; upper-wing appearing almost uniformly gray; outerwings quite translucent, with a narrow black trailing edge on undersurface; tail deeply forked, with dark edges. In breeding plumage, has black cap; bill and very short legs entirely red. Winter adult and immature have white forehead and black bill.

SIMILAR SPECIES. (1) Common Tern has darker wing tips; tail is less forked. Breeder has black-tipped red bill. (2) Forster's Tern has dark-tipped orange bill in breeding plumage, or black eye patch in winter. (3) Roseate Tern has pure white underwing and tail; mantle is entirely very light gray.

RANGE. Breeds in northern North America, Greenland, Iceland, and Eurasia; winters in high-latitude Southern Hemisphere. **Status**: Vagrant. One record: 20 Jun (1950). Matanzas harbor. **Habitat**: Oceanic. **Voice**: A harsh *tiar* or repeated *kee*. **Food**: Fish, offal, crustaceans.

FORSTER'S TERN

GAVIOTA DE FORSTER

Sterna forsteri   Pl. 22

DESCRIPTION. 14.5" (37 cm). White, with pale gray mantle, silvery white triangle on rear of wing, white rump, and deeply forked tail, with white edges. In breeding plumage, crown and nape black; bill orange with black tip. In nonbreeding plumage, head is white, with a black patch through eye; bill black, sometimes with red base. Juvenile resembles nonbreeding adult, with brownish crown and mottled brown above.

SIMILAR SPECIES. (1) Common Tern has broader dark wing tips. (2) Arctic Tern has grayer underparts and red bill. (3) Roseate Tern has pure white underwing, longer tail, and faster wingbeat. (4) Gull-billed Tern has tail only slightly forked, gray rump, and much heavier black bill.

RANGE. Breeds in southern Canada; western, central, and northeastern United States; and from northern Tamaulipas, Mexico east to southern Louisiana, United States. Winters south to Costa Rica, Bahamas, and Greater Antilles. **Status**: Rare winter resident in Cuba (30 Dec–Mar). **Habitat**: Freshwater marshes, ponds, ocean, flooded savannas, rice fields. **Voice**: A harsh nasal *kyar* and *keer*. **Food**: Fish, aquatic invertebrates, insects taken while flying over marshes.

LEAST TERN

GAVIOTICA; CINCELITO

Sterna antillarum   Pl. 22

DESCRIPTION. 9" (23 cm). Small, white, with pale gray mantle and black wing tips, moderately forked tail, and yellow legs. Breeder has black-tipped yellow bill and black cap with white patch on forehead. Juvenile has dusky yellow legs, black bill, brownish streaks on head, and dark postocular line.

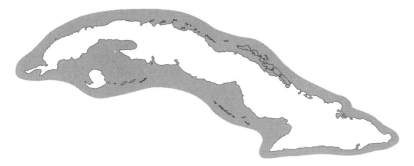

SIMILAR SPECIES. Winter adult and juvenile Black Tern have a white patch on hindneck; in flight, a black mark is visible on sides of upper chest; tail is short, only slightly forked; bill is dark.

RANGE. Breeds in United States, Honduras on Sandy Cay, and on islands off Venezuela; West Indies. Winters south to Brazil. **Status**: Common summer resident and transient in Cuba, Isla de Pinos, and many cays (18 Mar–Sep). **Habitat**: Open waters, sandy cays. **Nesting**: May–Jul. Lays up to four eggs, pale, tinted with olive or buff and spotted, blotched, and speckled in brown and blackish brown, in a small depression, near water. **Voice**: A shrill *peedee*. **Food**: Fish, aquatic invertebrates.

BRIDLED TERN
GAVIOTA MONJA
Sterna anaethetus   Pl. 23

DESCRIPTION. 15″ (38 cm). White below; dark gray above. Cap, eye stripe ("bridle"), and bill black; forehead white, extending behind eye; narrow white collar on hindneck; tail white, deeply forked; legs black. Juvenile light brown above, spotted with white, white below. Does not plunge into or alight on the surface of water as much as other terns.

SIMILAR SPECIES. Sooty Tern is all-black above; white on forehead does not extend behind eye.

RANGE. Breeds locally on tropical cays in all major oceans, migrating long distances, but remaining within warmer waters; West Indies. **Status:** Common, but local, summer resident on both coasts, most often seen offshore. Known to breed at Cayo Mono Grande, Cayo Felipe, Cayo los Ballenatos, Cayo Inglés. **Nesting:** May–Aug. Nests on bare sand or rock, sometimes in crevices. Lays one or two bluish-white eggs, spotted and speckled with brown and pale lilac. **Habitat:** Open waters, sandy cays. **Voice:** Harsh, barking cries, mostly at breeding sites. **Food:** Fish, squid.

SOOTY TERN
GAVIOTA MONJA PRIETA
Sterna fuscata   Pl. 23

DESCRIPTION. 16″ (41 cm). White below; black above. White on forehead extends rearward only to eye; cap, eye stripe, and bill are black; tail deeply forked, with white edges; legs black. Juvenile dark brown, the back covered with minute whitish spots; lower belly and undertail coverts grayish white. Like Bridled Tern, does not plunge into or alight on the surface of water as often as other terns.

SIMILAR SPECIES. Bridled Tern has white collar on hindneck; white on forehead extends behind eye. White on tail margins more extensive. Back is very dark gray rather than black. Wings appear slightly longer and narrower; flight is somewhat more graceful.

RANGE. Breeds locally on tropical cays in all major oceans, migrating long distances but remaining within warmer waters; in the Antilles off Cuba, in the Virgin Islands, and in the Lesser Antilles. **Status:** Locally common summer resident in Cuba. Known to breed at Mono Grande cay near Varadero, Felipes near Cayo Coco, and at los Ballenatos, Inglés, and de Dios cays in Archipiélago de los Canarreos. **Habitat:** Open waters. **Nesting:** May–Aug. Nests on bare sand or rock. Lays single white egg, spotted and speckled with reddish brown and pale lilac. **Voice:** Harsh barking cries, *wacky-wack*, often heard at night. **Food:** Fish, squid.

LARGE-BILLED TERN

GAVIOTA DE PICO LARGO

Phaetusa simplex   Pl. 23

DESCRIPTION. 15" (38 cm). White, with dark gray back. Upperwing has
bold tricolored pattern of black primaries, white secondaries, and gray innerwing. Bill heavy, long, and yellow; cap black; legs yellowish gray; and
tail short, slightly forked. Juvenile resembles adult but with brownish
spots, duller bill, and streaks on head.

SIMILAR SPECIES. None.

RANGE. Inland South America. **Status**: Vagrant. Two records: one undated;
28 May (1910). Harbors of Matanzas and Nipe. **Habitat**: Rivers, lakes,
coasts in the nonbreeding season. **Voice**: Silent. **Food**: Fish.

BLACK TERN

GAVIOTICA PRIETA

Chlidonias niger   Pl. 23

DESCRIPTION. 9.5" (24 cm). Nonbreeding adult white below, with a black
spot behind the eye; in flight, a black bar shows on sides of upper chest.
Breeding adults have black head, chest, and belly; white undertail coverts,
medium gray mantle and slightly notched tail; and black bill. Juvenile resembles winter adult, but with brownish back.

SIMILAR SPECIES. (1) Storm-Petrels are smaller, black, with hooked bill tip,
and a white band on rump. (2) Juvenile Least Tern is paler with deeply
forked white tail and contrasting black wing tips.

RANGE. Breeds in northern North America and Eurasia. In the Americas,
winters from Panama to Peru; West Indies. **Status**: Rare transient in Cuba
(25 Jul–21 Nov; 11 Apr–6 Jun). **Habitat**: Lagoons, rice fields, coasts.
Voice: A sharp, metallic *kick* or *krick*. **Food**: Fish, insects, mollusks.

BROWN NODDY

GAVIOTA BOBA; AURITA

Anous stolidus   Pl. 23

DESCRIPTION. 15" (38 cm). Dark brown, with white crown darkening to brown on nape. Tail wedge shaped and notched at tip. Bill and legs black. Juvenile has mostly brown head. Flight is strong and erratic.

SIMILAR SPECIES. None.

RANGE. Breeds locally in tropical islands worldwide. Wanders far at sea; West Indies. **Status**: Common summer resident in Cuba; nesting colonies at Mono Grande cay near Varadero, los Ballenatos, Inglés, and de Dios cays in Archipiélago de los Canarreos. **Habitat**: Open water near coasts. **Nesting**: May–Aug. Typically builds crude nest low in trees when present, but also on bare sand or rock. Lays two or three ashy-white eggs, spotted with reddish brown and pale lilac. **Voice**: A soft *kak*, also a *carrrk*. **Food**: Fish, squid.

BLACK SKIMMER

PICO DE TIJERA; GAVIOTA PICO DE TIJERA

Rynchops niger   Pl. 23

DESCRIPTION. 18" (46 cm). Winter plumage dark brown above with a white collar on lower neck and white underparts. Wings very long and narrow. Red, black-tipped bill is laterally compressed, with lower mandible longer than upper. Tail slightly forked. Entirely black above in breeding plumage. Juvenile mottled brown above. Feeds by flying low over the water, with tip of lower mandible cutting the surface. Swift flight.

SIMILAR SPECIES. None.

RANGE. Coasts of eastern United States, southern California, and Mexico, South America. Winters south to Argentina. **Status**: Uncommon winter resident and transient in Cuba and some cays (23 Nov–26 Apr). **Habitat**: Bays, inlets, lagoons, reservoirs. **Voice**: A barking *kau*. **Food**: Small fish, crustaceans.

AUKS Alcidae

Small to medium, somewhat penguin-like oceanic seabirds, mostly black above; white below. Rather heavy-bodied, with short bill, neck and legs. Wings short and narrow, wingbeats shallow and fast. Feed mainly by diving for marine invertebrates and fish, using the wings for underwater propulsion. Sexes alike. (**W**:24; **C**:1)

DOVEKIE

PINGÜINITO

Alle alle   Pl. 23

DESCRIPTION. 8" (20 cm). Small and stocky. Black above; white below. Chest may show brownish wash. In breeding plumage, head, throat, and breast black. Tail and bill very short.

SIMILAR SPECIES. None.

RANGE. Breeds on Arctic coasts, wintering in the North Atlantic. **Status**: Vagrant. Eight records: 12 Nov (1929); 11 Nov (1932); 21 Dec (1943); 9 Dec (1946); (1950s); 7 Dec (1962); two undated. Matanzas harbor, Gibara, Trinidad. **Habitat**: Oceanic. **Voice**: Silent outside the breeding season. **Food**: Primarily plankton.

PIGEONS AND DOVES Columbidae

Medium-sized, heavy-bodied birds with small heads and rather short wings and legs. Plumage is generally reddish brown, with inconspicuous markings visible mainly in flight, which is quite fast. Some have areas of iridescence on head or neck. They are commonly seen in pairs, and some species in flocks. Pigeons feed on the ground or in trees, mostly on grain and fruits. Nests are crude, and normally placed rather low in trees; a pair of white eggs is usually laid. Young are fed by regurgitation of a milky fluid. Sexes are generally alike, the females being somewhat duller. (**W**:310; **C**:13)

ROCK DOVE
PALOMA DOMÉSTICA
Columba livia   Pl. 24

DESCRIPTION. 12.5" (32 cm). Color variable from white through gray and brown to black. Ancestral color pattern is gray with white rump and black wing bars.

SIMILAR SPECIES. (1) White-crowned Pigeon is dark gray, with a white cap. (2) Scaly-naped Pigeon is dark gray all over; sides of neck are iridescent.

RANGE. Eurasia and northern Africa, east to southeast Asia. Introduced in virtually every inhabited part of the world. **Status**: Common throughout Cuba and Isla de Pinos. **Habitat**: Cities, villages, farms. **Nesting**: Mar–Sep. Builds crude nest in trees or buildings, laying two smooth, slightly glossy white eggs. **Voice**: Moaning *ooorh* or *oh-oo-oor*. **Food**: Grain, insects.

SCALY-NAPED PIGEON
TORCAZA CUELLIMORADA; TORCAZA MORADA
Columba squamosa   Pl. 24

DESCRIPTION. 15" (38 cm). Entirely dark gray, with a wine-red hood, iridescent on hindneck. Bare skin around eye is yellow to reddish orange. Juvenile has brown head and neck.

SIMILAR SPECIES. (1) White-crowned Pigeon has white or gray cap. (2) Plain Pigeon has reddish brown patch and white band on wing.

RANGE. Greater and Lesser Antilles. **Status**: Uncommon and very local permanent resident in eastern Cuba; Cayos Coco and Romano. Rather rare in

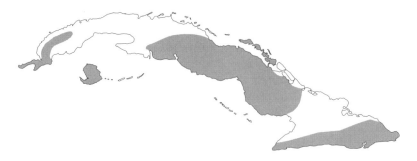

western Cuba and on Isla de Pinos. **Habitat**: Palm groves, areas with scattered tall trees, semideciduous woods, from sea level to rather high elevations. **Nesting**: Mar–Jun. A frail stick nest is made in base of palm fronds, or among bromeliad-crowded branches. Lays two glossy white eggs. **Voice**: Resembles that of White-crowned Pigeon, but more mournful: *oooo, OO-oo-oo, ROO-OO, . . . OO-oo-oo, ROO-OO,* etc. **Food**: Fruit, seeds, snails.

WHITE-CROWNED PIGEON

TORCAZA CABECIBLANCA
Columba leucocephala   Pl. 24

DESCRIPTION. 14" (36 cm). Entirely dark gray. Crown white in male, gray in female and immature. Lower neck has a touch of iridescent green, more obvious in male.

SIMILAR SPECIES. (1) Scaly-naped Pigeon has wine-red head and lacks white crown. (2) Plain Pigeon is brown, with white markings and reddish brown patch on wing.

RANGE. Southern Florida, islands of the western Caribbean Sea and West Indies. **Status**: Common permanent resident and transient in Cuba, Isla de Pinos, and several cays. Found in large numbers where fruit is abundant. **Habitat**: Semideciduous woods, mangroves. Commonly commutes long distances, even over water, in early morning and late afternoon, from roosts or breeding colonies to feeding areas. **Nesting**: Apr–Aug. Nests in colonies as large as many hundreds of pairs, generally in mangrove cays, also in woodlands. Builds a platform of twigs, roots and plant-stems. Usually lays two white eggs. **Voice**: A grunting, somewhat hoarse *oooo, OO-oo, OOOO, . . . OO-oo, OOOO*, etc. **Food**: Fruit, seeds, occasionally insects.

PLAIN PIGEON
TORCAZA BOBA; PALOMA BOBA
Columba inornata   Pl. 24

DESCRIPTION. 15 " (38 cm). Brown, with gray wings, rump, and tail. Belly is vinaceous. Upper surface of inner wing has a thin white median band and a reddish brown patch. Eye white; eye ring red. Juvenile duller with dark eye.

SIMILAR SPECIES. (1) Scaly-naped Pigeon has uniformly gray wing and wine-red head and neck. (2) White-crowned Pigeon has white or gray cap. (3) White-winged Dove is smaller, with tail tip marked with black and white, and conspicuous white wing patches.

RANGE. Greater Antilles. **Status**: An endangered permanent resident on Cuba and Isla de Pinos. Remaining populations inhabit the coastal areas of the Guanahacabibes Peninsula; environs of Maneadero, Zapata peninsula; and Sierra de Najasa, Camagüey province. **Habitat**: Coastal and semideciduous woods, low forested hills, grassy palm groves. **Nesting**: Apr–Jul. Builds a fragile stick nest lined with grass. Lays two white eggs. **Voice**: A guttural call, *hoowua, HOO-hoowua, . . . HOO-hoowua*, etc. **Food**: Fruit, seeds.

EURASIAN COLLARED-DOVE
TÓRTOLA DE COLLAR
Streptopelia decaocto   Pl. 51

DESCRIPTION. 12 " (30 cm). Entirely pale grayish brown, with an incomplete black collar. Tail long and square with slightly rounded corners, tipped grayish white. Gray undertail coverts. Underside of tail black with strongly contrasting white distal half.

SIMILAR SPECIES. None.

RANGE. Throughout Eurasia. Introduced in the Bahamas in 1974, and has subsequently invaded Florida, Cuba, and much of Lesser Antilles. **Status**: Confirmed in Cuba in 1990, but unconfirmed sightings from 1988. Observed in La Habana, Guanahacabibes peninsula, and Cayo Coco. Spreading rapidly and increasing in numbers. **Habitat**: Urban areas. **Nesting**: Mar–Aug. The nest is an unlined platform of twigs constructed in a bush or tree. Lays two white eggs. **Voice**: A rather soft, repeated *kuuk-kooooooo-kuuk*. **Food**: Fruit, grain.

WHITE-WINGED DOVE

PALOMA ALIBLANCA

Zenaida asiatica   Pl. 25

DESCRIPTION. 11" (28 cm). Grayish brown. A white wing patch, very conspicuous in flight, noticeable on the folded wing, where it forms a white border, particularly visible along the leading edge. A black crescent is visible under cheek. Sides of lower neck have iridescent purple sheen, usually more conspicuous in male. Tail feathers, except central ones, broadly tipped white. Large flocks form where food is plentiful. Juvenile lacks iridescence.

SIMILAR SPECIES. (1) Zenaida Dove is cinnamon brown, with white on wing restricted to trailing edge of secondaries. (2) Mourning Dove has plain brownish gray wing and long pointed tail. (3) Plain Pigeon is larger, with unmarked tail.

RANGE. Southern United States to Chile; Bahamas and Greater Antilles. **Status**: Common permanent resident in eastern and central Cuba, less so in western provinces, Isla de Pinos, and some larger cays. **Habitat**: Semideciduous and pine forests, wooded hills, mangrove swamps. **Nesting**: Apr–Jul. Nest singly or in loose groups in a tree or shrub. Builds a shallow, thin platform of twigs and grasses. Lays two white eggs. **Voice**: Perhaps the loudest among pigeons: *ooh-woo-woo-woo*. Two common rhythms are *ooah-AHoo* and *who cooks for you all*. **Food**: Fruit, seeds, grain.

ZENAIDA DOVE
GUANARO; PALOMA SANJUANERA
Zenaida aurita   Pl. 24

DESCRIPTION. 11" (28 cm). Cinnamon brown, paler below. Sides of lower
neck iridescent purple, usually more conspicuous in male. Two black spots
behind cheek. Secondaries and all but central tail feathers have white tips;
tail rounded. Bill black. Overall, males average browner above than females, which tend to be gray brown, and have larger black spots and more
extensive white on wing. Breast has darker vinaceous tint. Juvenile duller.
SIMILAR SPECIES. (1) White-winged Dove is grayish brown, with a bold
white band on wing. (2) Mourning Dove has long, pointed tail and lacks
pale tips to secondaries.
RANGE. West Indies, north coast of Yucatán peninsula. **Status**: Common
permanent resident in Cuba, Isla de Pinos, and some of the larger cays.
Habitat: Forest, coastal vegetation, pine woods, and clearings, where
flocks gather to feed. **Nesting**: Mar–Jul. Builds a thin platform of twigs in
a bush or tree. Lays two white eggs. **Voice**: A clear *OOLA, OO, OO-OO*,
with a rather sharp second syllable. Very similar to Mourning Dove, but
slightly deeper and faster. **Food**: Fruit, seeds, grain.

MOURNING DOVE
PALOMA RABICHE
Zenaida macroura   Pl. 24

DESCRIPTION. 12" (30 cm). Grayish brown above, paler below. Sides of lower neck iridescent purple, usually more evident on male. Tail long and pointed, bordered black and white. Single black auricular spot. Sometimes form large flocks where food is abundant. Juvenile darker, spotted with black on back, chest, and wings. Flight nervous and quick, with irregular wingbeats.

SIMILAR SPECIES. Both (1) Zenaida and (2) White-winged Doves have shorter, rounded tail and the latter species has a bold white wing patch.

RANGE. Southern Canada, United States, Mexico, Costa Rica, Panama, Bahamas, and Greater Antilles. **Status**: Abundant permanent resident in Cuba and Isla de Pinos; less so on the larger cays. Also a regular winter resident. **Habitat**: Savannas, pastures, rice fields, coastal vegetation, cities and towns. **Nesting**: Feb–Sep. Builds a nest of twigs and grasses at a low to medium elevation in a bush or tree. Lays two white eggs. **Voice**: A lengthy and mournful *OOWA, OO, OO-OO*. Very similar to Zenaida Dove, but slightly more drawn out, breathy, and higher pitched; second syllable not as clipped. **Food**: Seeds, grain, some fruit and snails.

PASSENGER PIGEON

PALOMA MIGRATORIA

Ectopistes migratorius   Pl. 26

DESCRIPTION. 8.6" (23 cm). Very similar in appearance to Mourning Dove, but larger with darker, slate gray upperparts; underparts with pink violaceous tinge. Tail long and pointed. Female smaller and duller than male.

RANGE. Bred formerly from central Montana, east central Saskatchewan, southern Manitoba, Minnesota, Wisconsin, Michigan, Ontario, southern Quebec, New Brunswick, and Nova Scotia south to eastern Kansas, Oklahoma, Mississippi, and Georgia. **Status**: Extinct. Vagrant. Four records from last century. **Habitat**: Open country and cultivated lands adjacent to forest. **Food**: Nuts, seeds, grain, fruit.

COMMON GROUND-DOVE

TOJOSA; TOJOSITA

Columbina passerina   Pl. 25

DESCRIPTION. 6.5" (17 cm). Very small, grayish brown, with faintly scaled head and breast and scattered, small dark-violet spots on wings. In flight, outer wing is rufous and the tail short and rounded, with black edges and white corners. Male has bluish crown and nape, and pink flush on breast. Female is duller, plain gray. Juvenile similar to female, with cinnamon brown feather tips on back and wings. Spends most of the time on the ground, occasionally perching in trees. Flies only short distances when disturbed.

SIMILAR SPECIES. None.

RANGE. Southern United States to northern South America (eastern Brazil); West Indies. **Status**: Common permanent resident in Cuba, Isla de Pinos, and many cays. **Habitat**: Broad forest trails, coastal vegetation, open country, towns. **Nesting**: Jan–Jul. May nest on ground or on branches. Builds a thin frail platform of fine twigs, grasses, stems, rootlets, and occasionally feathers. Lays two white eggs. **Voice**: A soft, repeated *wah-up*, repeated in long series. **Food**: Seeds, grain, insects, fruit.

KEY WEST QUAIL-DOVE
BARBIQUEJO
Geotrygon chrysia   Pl. 25

DESCRIPTION. 11" (28 cm). Reddish brown above, whitish below. Breast pale vinaceous. Purple-green iridescence on head and back, and rufous of wings and tail brighter and more extensive in male. Females duller and less iridescent. A conspicuous white stripe under eye. Juvenile is duller with dark borders on wing coverts. Usually seen on ground, but will perch low in trees. Flight erratic.

SIMILAR SPECIES. (1) Male Ruddy Quail-Dove is smaller with reddish brown back, pinkish brown underparts, and a buff stripe under eye; female is olivaceous brown. (2) Blue-headed Quail-Dove has blue cap and black patch on throat and breast.

RANGE. Bahamas, Cuba, Hispaniola (including Gonâve, Tortue, and Catalina islands), Puerto Rico, and Vieques Island. **Status**: Uncommon to locally common permanent resident on Cuba, Isla de Pinos, and some of the larger forested cays. **Habitat**: Dry forest with little undergrowth, mainly near the coast. **Nesting**: Year-round. Builds nest near the ground or low in trees, laying two beige eggs. **Voice**: A very deep *ooowooo*, sliding slightly downward in pitch and repeated in series at about 3-second intervals. It is quite similar to Ruddy Quail-Dove, but slower, more mournful. **Food**: Fruit, seeds, small snails.

GRAY-HEADED QUAIL-DOVE

CAMAO; AZULONA

Geotrygon caniceps   Pl. 25

DESCRIPTION. 11" (28 cm). Dark gray above, gray below. Male has an overall conspicuous blue and purple iridescence, with white forehead and a pale gray crown. Female grayer on head and with less iridescence on body. Juvenile brownish. After perching, or while walking, bobs the body.

SIMILAR SPECIES. None.

RANGE. Cuba and Dominican Republic. **Status**: Rare and local permanent resident on Cuba, although common in Zapata peninsula and Sierra del Rosario. Vulnerable. **Habitat**: Forest, near swamps or at medium elevations. Occurs in highlands of Dominican Republic and may thus be expected at higher elevations in Cuba. **Nesting**: Jan–Aug. Builds a loose nest of twigs and leaves, 1–3 m above the ground, on the main trunk or a branch. Usually lays a single cream-colored egg, rarely two. **Voice**: A low, rapidly repeated *uup-uup-uup* in a long series, and a more measured, louder and ascending *uoop*. **Food**: Fruit, seeds.

RUDDY QUAIL-DOVE

BOYERO; TORITO; CAMAO

Geotrygon montana   Pl. 25

DESCRIPTION. 10" (25 cm). *Male*: reddish brown above, with back washed with purple; a buff stripe below the eye. Underparts vinaceous buff. *Female*: olivaceous brown above; beige brown below, except for white

throat. Facial stripe less distinct than in male. Juvenile similar to female; back feathers tipped with ochre. Usually perches low, and like other quail-doves is most often found on the ground. Flight low and very fast. Flushes more readily than other quail-doves.

SIMILAR SPECIES. Key West Quail-Dove is larger with green and purple iridescence on head and back and whitish underparts.

RANGE. Central America south to northern Argentina; West Indies. **Status**: Common permanent resident in Cuba, Isla de Pinos, and some larger cays on the northern coast. **Habitat**: Dense forest, from sea level to medium elevations; coffee plantations. **Nesting**: Feb–Jul. Builds a loose nest of twigs and leaves 1–6 m above ground. Lays two cream-colored eggs. **Voice**: A simple, deep *hooo* with no internal variation in pitch, repeated in series at about 3-second intervals. **Food**: Seeds, fruit, small snails.

BLUE-HEADED QUAIL-DOVE
PALOMA PERDIZ
Starnoenas cyanocephala   Pl. 25

DESCRIPTION. 12" (30 cm). Cinnamon brown, with an olive wash. Throat and breast black, flecked with iridescent blue and surrounded with white. Cap metallic blue; sides of neck streaked. A long, conspicuous white stripe under eye. Juvenile similar, with dark-tipped cap feathers. Wing sound is very loud when flushed.

SIMILAR SPECIES. Key West Quail-Dove has green and purple iridescence on head and pale vinaceous breast.

RANGE. Endemic to Cuba. **Status**: Rare and vulnerable. Virtually extirpated on Isla de Pinos. Uncommon; only found regularly on Guanahacabibes peninsula, at La Güira, and in the Zapata peninsula. **Habitat**: Thick forest with little undergrowth. **Nesting**: Apr–Jun. Builds a loose nest of twigs, above bromeliads or on trunk, near the ground. Lays two white eggs. **Voice**: A deep two-syllable * oooowup . . . oooowup*, repeated in long series at about 2-second intervals, each note ending abruptly. **Food**: Fruit, seeds.

PARAKEETS AND PARROTS Psittacidae

Small to large birds with heavy bodies, large heads, and strong curved bills. Feet are strong, with two backward-pointing toes. Mostly tropical, these birds are colorful, noisy, gregarious, and capable of remarkable gymnastics while perching and feeding, hanging by one foot, for example. They feed on a variety of seeds and fruits. Their wingbeats are shallow and below the horizontal. (**W**:358; **C**:2)

CUBAN MACAW
GUACAMAYO
Ara tricolor   Pl. 26

DESCRIPTION. A beautiful tricolored bird with a mainly red body, blue wings and tail, and a distinctive yellow nape.

RANGE. Cuba and Isla de Pinos. **Status**: Extinct. Last specimen taken at Zapata peninsula in 1864. **Habitat**: Forest edge and open country with scattered trees, especially palms. **Nesting**: Probably in dead palms. **Voice**: Undescribed. **Food**: Seeds, fruits.

CUBAN PARAKEET
CATEY; PERIQUITO
Aratinga euops   Pl. 27

DESCRIPTION. 10" (25 cm). Vivid green, with scattered red spots on head, sides of neck, and breast. A large red area on underwing at bend of wing. Tail long and pointed. Juvenile duller.

SIMILAR SPECIES. None.

RANGE. Endemic to Cuba, and formerly Isla de Pinos. **Status**: Vulnerable. Restricted to Zapata peninsula, the mountains of Trinidad, Peralejo, Guasimal, Loma de Cunagua, Sierra de Najasa, some areas of the eastern mountain ranges. **Habitat**: Intact or undisturbed forest, savannas with palm groves. **Nesting**: May–Aug. Lays two to five white eggs in abandoned woodpecker holes in palms or other tree cavities, and occasionally in potholes in cliffs. **Voice**: A repeated squeak *crik-crik-crik*, mostly in flight. A low murmur while perched. **Food**: Seeds, fruits.

CUBAN PARROT

COTORRA; PERICO

Amazona leucocephala   Pl. 27

DESCRIPTION. 12" (30 cm). Green, with red cheeks, throat, and chest. Forehead and bare skin around eye white. Primaries blue; undertail and adjacent coverts yellow, green, and red. Bill and legs yellow. Juvenile is duller. Wingbeats shallow and rapid.

SIMILAR SPECIES. None.

RANGE. Bahamas (Great Inagua and Abaco), Cuba, Cayman Islands (Grand Cayman and Cayman Brac). **Status**: A permanent resident on Cuba and Isla de Pinos. Rare and vulnerable; today restricted to Guanahacabibes peninsula, Zapata peninsula (where it is still common), southern and western Isla de Pinos, Macizo de Guamuhaya, Loma de Cunagua, Sierra de Najasa, and the forests of western Sierra Maestra and Cuchillas del Toa. **Habitat**: Well-preserved forest with mature trees and snags, savannas with palm groves. **Nesting**: Mar–Jul. Lays three or four white eggs in abandoned woodpecker holes, usually in palms. **Voice**: A very loud repeated *squack-squack*, especially when taking flight. **Food**: Fruits, seeds.

CUCKOOS AND ANIS Cuculidae

Medium-sized, mostly dull-colored, solitary birds with short wide wings and long tails. Bills are rather long and slightly decurved. Feet have parrot-like toe arrangement—two forward, two backward—an adaptation for climbing and hopping. Calls are loud. Nests are crude; eggs are green or blue. Mainly insectivorous. Sexes alike. (**W**:171; **C**:5)

BLACK-BILLED CUCKOO
PRIMAVERA DE PICO NEGRO
Coccyzus erythropthalmus   Pl. 27

DESCRIPTION. 12" (30 cm). Grayish brown above; white or faintly buffy below. Tail feathers have narrow white tips. Bill slightly decurved and entirely black. Bare skin around eye orange red. Juvenile has buffy eye ring, and may show some rufous in primaries.

SIMILAR SPECIES. (1) Yellow-billed Cuckoo has yellow lower mandible with black tip, and tail feathers that are more broadly tipped white. In flight, shows a bright rufous patch on primaries. (2) Mangrove Cuckoo has black patch behind eye, belly buff, and yellow at base of lower mandible.

RANGE. Eastern and central North America, wintering from northern South America to Peru and central Bolivia. **Status**: Very rare fall transient in Cuba; only one spring record (15 Sep–15 Nov; 11 May). **Habitat**: Coastal vegetation and scattered trees. **Voice**: A low, hollow *cu-cu, cu-cu-cu* or *cu-cu-cu-cu*, repeated rhythmically in long series, mostly on breeding grounds. **Food**: Insects, mollusks, fruits.

YELLOW-BILLED CUCKOO
PRIMAVERA; PRIMAVERA DE PICO AMARILLO
Coccyzus americanus   Pl. 27

DESCRIPTION. 12" (30 cm). Grayish brown above; white below. Tail feathers black with broad white tips, conspicuous when seen from below. Outer wings rufous above. Bill slightly decurved; lower mandible yellow with black tip. Juvenile has paler undertail pattern, sometimes without yellow on lower mandible.

SIMILAR SPECIES. (1) Black-billed Cuckoo has entirely black bill and gray undertail with narrow, pale tips. (2) Mangrove Cuckoo has belly buff, black patch behind eye, markedly decurved bill, and wings entirely grayish brown above, without rufous.

RANGE. Southeastern Canada, United States, Mexico, and Greater Antilles, wintering from northern South America to northern Argentina; West Indies. **Status**: Uncommon summer resident in Cuba (10 Feb–Nov), though few breeding records. Transients have been observed and collected in Oct–Nov at La Habana Botanical Garden and also Cayos Coco and Paredón Grande. **Habitat**: Forest, wooded areas, shrubs. **Nesting**: Apr–Jul. Builds a shallow twig platform, with rootlets, dry leaves, and grasses, in a bush or a tree. Lays four or five greenish-blue eggs. Occasionally parasitizes nests of other birds. **Voice**: A fast guttural cackling diminishing in loudness *ka-ka-ka-ka-ka-ka, kow-kow-kow*. **Food**: Insects, caterpillars, lizards.

MANGROVE CUCKOO
ARRIERITO; PRIMAVERA

Coccyzus minor   Pl. 27

DESCRIPTION. 12" (30 cm). Grayish brown above, with white throat and breast and buff belly. A black patch behind the eye. Tail feathers black with broad white tips, conspicuous when seen from below, as in Yellow-billed Cuckoo. Bill strongly decurved; base of lower mandible yellow. Juvenile has duller patch behind eye and tail pattern.

SIMILAR SPECIES. (1) Yellow-billed and (2) Black-billed Cuckoos are entirely white below and lack black patch behind eye.

RANGE. South Florida, Mexico to northern South America; West Indies; winters throughout the breeding range. **Status**: A locally common permanent resident on larger cays such as Francés, Santa María, Coco, Paredón Grande, Romano, Cantiles, and Cayo Largo. Less common on Cuba and not known from Isla de Pinos. **Habitat**: Mangroves, low coastal scrub.

Nesting: Apr–Jul. Builds a flimsy stick nest low on branches, laying three greenish-blue eggs. **Voice**: A low, guttural *ga-ga-ga, ga-ga, gau-gau*. **Food**: Mainly insectivorous: caterpillars, crickets, butterflies, stick insects; spiders.

GREAT LIZARD-CUCKOO
ARRIERO; GUACAICA; TACÓ
Saurothera merlini   Pl. 27

DESCRIPTION. 20″ (51 cm). Large. Olive brown above; gray and buff below. Tail very long, undertail has a bold pattern of black and white. Bill very long, slightly decurved, and blue gray. Bare skin around eye orange red. Three races have been described for the Cuban archipelago. Those from cays north of central Cuba and from Isla de Pinos have paler and more extensive gray areas on underparts, and are smaller than the main-island race. Juveniles of all races have paler tail pattern and yellow bare skin around eye. This poor flier commonly glides from treetops and often forages on ground. Runs frequently.

SIMILAR SPECIES. All other cuckoo species are considerably smaller, with shorter tails and bills, and white breasts.

RANGE. Bahamas and Cuba. **Status**: Common permanent resident on Cuba, Isla de Pinos, and the larger forested cays north of Camagüey and Ciego de Ávila provinces. **Habitat**: Forest, wooded areas, shrubs, from sea level to fairly high elevations. **Nesting**: Apr–Oct. Builds a flimsy stick nest with leaves, at low or middle levels in trees or bushes, laying two or three white eggs. **Voice**: A two-part call, *tacooo-tacooo, ka-ka-ka-ka-ka-ka*, the second part gradually becoming louder. Also a guttural *tuc-wuuuh*. **Food**: Small vertebrates: lizards, frogs, snakes, bird eggs and nestlings, large insects.

SMOOTH-BILLED ANI
JUDÍO; GARRAPATERO
Crotophaga ani   Pl. 27

DESCRIPTION. 14.5″ (37 cm). Entirely black with some iridescence. Bill very deep, rather short, and much compressed laterally. The very long tail

moves so loosely that it seems only pinned to rear of the body. Immature is duller. Flight slow, with alternate flapping and gliding on short, flat wings.
SIMILAR SPECIES. None.
RANGE. Florida, Costa Rica to northern Argentina; West Indies. **Status**: Abundant permanent resident in Cuba, Isla de Pinos, and several cays. **Habitat**: Grassy or brushy areas, pastures, road sides. **Nesting**: Apr–Oct. Nests are large, communal, and hidden in thick vegetation. Several females may lay up to 20 greenish-blue eggs in one nest. **Voice**: A loud, repeated rising *weee-iuk*. **Food**: Lizards; insects; especially grasshoppers and parasites of cattle; fruit; seeds.

BARN OWLS Tytonidae

Nocturnal raptors with large head and striking, heart-shaped faces. The eyes are rather small and face forward, and legs are long. Nests are situated in natural cavities, in abandoned buildings, or on the ground in secluded places. Prey is swallowed whole, and a large pellet with indigestible parts is later regurgitated. Females are somewhat larger than males. (**W**:17; **C**:1)

BARN OWL
LECHUZA
Tyto alba   Pl. 28

DESCRIPTION. 15" (38 cm). Yellowish-cinnamon brown above, the back mottled with gray; white to pale beige below, with sparse and minute brown spots. At night, flying birds appear entirely white below. Often perches on electric wires.

SIMILAR SPECIES. None.

RANGE. Cosmopolitan in tropical and temperate regions. **Status**: Common permanent resident in Cuba, Isla de Pinos, and some cays; very rare winter resident. **Habitat**: Open country, settled areas, including towns and cities, cane fields. **Nesting**: Year-round. Lays two or three white eggs in natural tree cavities, caves, or abandoned buildings. **Voice**: In flight, a raspy hissing scream *shhh*. During display flight, rapid, almost continuous, clicking. **Food**: Principally rodents; also reptiles and small birds.

TYPICAL OWLS Strigidae

Small to large nocturnal raptors generally mottled with brown tones. Head is large, with large forward-facing eyes in the rounded face, from which feathers grow outward, forming facial disks. Ear tufts are often present. Nests are placed in natural cavities, abandoned woodpecker holes, abandoned hawk nests, and on ground. Prey is swallowed whole and a pellet with indigestible parts is later regurgitated. Females are somewhat larger than males. (**W**:161; **C**:6)

CUBAN SCREECH-OWL
SIJÚ COTUNTO; CUCO
Otus lawrencii   Pl. 28

DESCRIPTION. 8.5" (22 cm). Dark brown above, mottled with white; pale grayish white below, with faintly streaked chest and belly. Conspicuous white eyebrow; brown eyes. Legs rather long and unfeathered. Strictly nocturnal.

SIMILAR SPECIES. (1) Burrowing Owl is barred below, with yellow eyes and a white stripe across chin. (2) Cuban Pygmy-Owl is smaller, mottled below, with yellow eyes and entirely feathered legs.

RANGE. Endemic to Cuba. **Status**: Common on Cuba, Isla de Pinos, Cayos Coco and Romano. In order to be consistent with AOU (1998), we have followed their unexplained systematic treatment of this species. However, we believe this species is best retained within the monotypic genus, *Gymnoglaux*. **Habitat**: Well-preserved dense forest. **Nesting**: Mar–May. Natural cracks and holes in trees; abandoned woodpecker holes; usually lays two white eggs. **Voice**: A series of 12 to 15 low *cu* notes, rapidly accelerating in the manner of a bouncing ball. Pairs sing antiphonally on slightly different pitches. Also a repeated *wheer* or *wheep*, rising and then falling. **Food**: Small reptiles, large insects.

CUBAN PYGMY-OWL

SIJÚ PLATANERO; SIJUCITO; SIJÚ

Glaucidium siju   Pl. 28

DESCRIPTION. 6.5" (17 cm). Small and compact. Brown above, with whitish bars on back and a tawny collar; white mottled with brown below. The short legs are entirely feathered. Eyes yellow; tail very short. Juvenile has unmarked back and streaked belly. At times, cocks tail while perched, and jerkily moves it from side to side. A diurnal and nocturnal hunter; normally easy to approach.

SIMILAR SPECIES. Cuban Screech-Owl is larger and streaked below with brown eyes and bare legs.

RANGE. Endemic to Cuba. **Status**: Common on Cuba, Isla de Pinos, and Cayo Romano north of Camagüey and Ciego de Ávila; Cayo Cantiles. **Habitat**: Forest of all kinds, from sea level to high mountains. **Nesting**: Mar–May. Nests in natural cavities or abandoned woodpecker holes. Lays three or four white eggs. **Voice**: A single, rather high-pitched *too* at intervals; an accelerating series of *kew* notes, rising in pitch and changing in quality, more prolonged and strident in males, could be confused with Osprey. Also a high-pitched *wheee*, heard commonly during courtship. **Food**: Lizards, insects, small birds.

BURROWING OWL
SIJÚ DE SABANA; CUCO DE SABANA; CUZCO
Athene cunicularia Pl. 28

DESCRIPTION. 9″ (23 cm). Dark brown above, extensively and conspicuously spotted with white. Eyes yellow. Breast and belly whitish, barred with brown; a white stripe across chin. Legs very long; tail rather short. Juvenile buff below. Terrestrial, hunting by hovering over open country, frequently in daylight. Frequently bobs up and down when agitated.

SIMILAR SPECIES. Cuban Screech-Owl is similar in size but is a strictly nocturnal forest species, with brown eyes and faint streaks below.

RANGE. Western and central North America and Florida, central Mexico. Locally in South America, Bahama, Cuba, Hispaniola (including Gonâve and Beata islands), and northern Lesser Antilles. **Status**: Rare and local permanent and winter resident in Cuba: La Fé and Cortés in Pinar del Río province, Los Indios on Isla de Pinos, north of Itabo in Matanzas province, Jauco and Jamaica in Guantánamo; also recorded on some of the larger cays north of Ciego de Ávila and Camagüey provinces. **Habitat**: Grassy-sandy savannas, old fields. **Nesting**: May–Aug. Lays five to eight round white eggs at the end of excavated tunnels up to 2 m long. **Voice**: A rapid, chattering *quick-quick*. **Food**: Small reptiles, large insects.

LONG-EARED OWL
BÚHO
Asio otus Pl. 28

DESCRIPTION. 15″ (38 cm). Dark brown above, spotted with darker brown; cream-colored chest and belly, streaked with dark brown. Eye yellow, facial disk tawny brown. Two very long ear tufts. Long wings may extend beyond tail and show a buff patch above and black wrist mark below.

SIMILAR SPECIES. (1) Short-eared Owl has bolder wing pattern, very short ear tufts, and pale buff facial disks. Found in open country. (2) Stygian Owl is bulkier, with shorter ear tufts and darker brown facial disk.

RANGE. Breeds in northern United States and Canada, wintering south to Mexico. Also Europe and Asia. **Status**: Vagrant. One record. La Habana. **Habitat**: Forest. **Voice**: Common call is one or more long *hooo*'s. **Food**: Rodents, birds, fish, reptiles, insects.

STYGIAN OWL
SIGUAPA

Asio stygius   Pl. 28

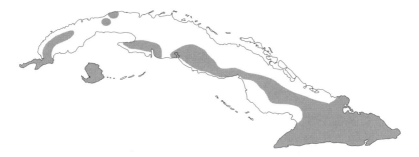

DESCRIPTION. 17" (43 cm). Dark brown, with cream-colored and white spots above; streaked below. Two rather long ear tufts. Facial disk brown. Eyes reddish orange. Juvenile has barred belly.

SIMILAR SPECIES. (1) Very rare Long-eared Owl is slimmer, with longer ear tufts and tawny brown facial disk. (2) Short-eared Owl is smaller, has barely visible ear tufts, and pale buff facial disk, and is found in open country.

RANGE. Mexico, Central and South America; Cuba, Hispaniola, and Gonâve. **Status**: Uncommon, local and vulnerable permanent resident on Cuba and Isla de Pinos: La Güira, Sierra de la Güira, Pinar del Río province; Zapata peninsula; Sierra de Guamuhaya, Cienfuegos, and Sancti-Spíritus provinces; and Sierra Maestra. **Habitat**: Semideciduous and pine forests. **Nesting**: Jan–Apr. Nests high in trees and also in holes in trees. May add sticks to an old hawk nest or build a platform with sticks. Lays two white eggs. **Voice**: Male, a short, hushed, *hooo*; female, *quick-quick*. **Food**: Birds, bats, reptiles.

SHORT-EARED OWL
CÁRABO

Asio flammeus   Pl. 28

DESCRIPTION. 15" (38 cm). Dark brown, mottled with buff spots, and streaked. Upper breast boldly streaked with dark brown. Eyes yellow, surrounded by black feathers, with contrasting white feathering around the bill. Two short ear tufts. Hunts mainly at dawn and dusk. In flight, shows a conspicuous black patch at wrist and a large buff area on upperwing. Flight is erratic and low, with wings angled upward.

SIMILAR SPECIES. (1) Stygian Owl is bulkier with much more apparent ear tufts and is a forest species. (2) Long-eared Owl has less prominent wing markings, long ear tufts, and reddish facial disk.

RANGE. Eurasia; Hawaiian Islands, Caroline Islands, northern United States and Canada. American populations winter south to Mexico; Greater Antilles. **Status**: Common permanent resident in Cuba, increasing in numbers. Not known from Isla de Pinos. **Habitat**: Rice fields, savannas, pastures, citrus plantations. **Nesting**: Probably whole year. Builds crude nest on the ground. Lays three or four white eggs. **Voice**: Male produces a rapid series of 16 or more short, low notes *uh-uh -uh-uh*; female, a sort of *yip*. **Food**: Rodents, reptiles, large insects.

GOATSUCKERS Caprimulgidae

Small to medium-sized birds, with very small bill and very short legs. Large eyes reflect light at night. Gape is enormous, though not evident in birds seen at rest in daytime, with a few long bristles protruding forward around mouth, to aid in capturing insects on the wing. Tail is long. Plumage is highly cryptic, mottled browns with intricate patterns. Most species are nocturnal; some are crepuscular, spending the day perching motionless on a branch or on the ground. Nighthawks hunt for and capture prey during flight, which is often bounding. Nightjars generally capture prey by sallying from a perch or the ground. No nest is built; the eggs are laid directly on the ground. (**W**:76; **C**:5)

COMMON NIGHTHAWK
QUEREQUETÉ AMERICANO
Chordeiles minor

Not Illustrated

DESCRIPTION. 9.5" (24 cm). Mottled dark grayish brown, with strongly barred underparts. Wings long and pointed, with a conspicuous white band across the primaries. Tail slightly forked, barred black and white.

Male has white throat and a conspicuous white subterminal tail band. Female has buff throat and lacks white tail band. Juvenile similar to female but paler. Bounding flight consists of deep wingbeats interspersed with long glides.

SIMILAR SPECIES. Antillean Nighthawk is very similar, but wing linings and underparts have more buff. Best differentiated by call. Glides between wingbeats shorter.

RANGE. Breeds in North and Central America, wintering in South America. **Status**: Uncommon transient in Cuba and Isla de Pinos (27 Aug–7 Oct; Apr–10 May). **Habitat**: Open country. **Voice**: A resonant, nasal *peeent*, given in flight, often in darkness. **Food**: Insects.

ANTILLEAN NIGHTHAWK

QUEREQUETÉ

Chordeiles gundlachii Pl. 29

DESCRIPTION. 9.5" (24 cm). Mottled dark grayish brown, with underparts strongly barred and suffused with buff. Wings long and pointed, with a very conspicuous white band across the primaries. Wing linings buff, barred with black. Tail slightly forked, barred in black and white. Male has white throat and a conspicuous white tail band. Female has buff throat and lacks tail band. Juvenile similar to female but paler. Commonly seen and heard at dawn and dusk, when it displays by diving from considerable heights and producing a booming sound. Bounding flight consists of deep wingbeats interspersed with glides.

SIMILAR SPECIES. Common Nighthawk is virtually identical, but lacks buff wing linings and underparts. Best differentiated by call. Glides are shorter than those of Common Nighthawk.

RANGE. Breeds in West Indies and Florida Keys; presumed to winter in South America. **Status**: Common summer resident and transient in Cuba, Isla de Pinos, and some larger cays (6 Feb–Oct). **Habitat**: Open forest, savannas, pastures. **Nesting**: Apr–Jul. Lays single (rarely two) bluish-brown spotted egg on bare ground. **Voice**: In flight, gives a sharp and clear *ke-re-ke-te* with slight accent on last syllable, often heard at night and on cloudy days. **Food**: Insects.

CHUCK-WILL'S-WIDOW

GUABAIRO AMERICANO

Caprimulgus carolinensis   Pl. 29

DESCRIPTION. 12" (30 cm). Mottled brown all over; paler below. Male has white bar at base of throat and three outermost tail feathers with white inner webs. Female has buff bar at base of throat. Forages for insects close to the ground. Tail and wings rounded. Flight fast, alternating wingbeats with glides.

SIMILAR SPECIES. (1) Greater Antillean Nightjar is smaller, less warmly toned, with breast heavily spotted with white. Male's outer tail feathers tipped white. At close range, longer mouth bristles are conspicuous. (2) Whip-poor-will is smaller and grayer, with black throat and U-shaped necklace; outer tail feathers have extensive white markings.

RANGE. Eastern North America, wintering to Colombia; Bahamas and Greater Antilles. **Status**: Uncommon winter resident in Cuba, Isla de Pinos, and some larger cays. Also a common transient (28 Aug–28 May). **Habitat**: Clearings in forest, fields, dry brushy terrain, wetlands. **Voice**: A four-syllable *chuk-will-wi-dow*, the first inaudible at a distance. Also a croaking sound given in flight. **Food**: Insects.

GREATER ANTILLEAN NIGHTJAR

GUABAIRO

Caprimulgus cubanensis   Pl. 29

DESCRIPTION. 11" (28 cm). Mottled grayish brown all over; breast and belly blackish brown, spotted with white. Bar at base of throat buffy. Bristles very long and curved inward. Tail and wings rounded. Male has conspicuous white tips on outer tail feathers, buffy in female. Regularly seen at night "jumping" from the ground to capture low-flying insects.

SIMILAR SPECIES. (1) Chuck-will's-widow is larger, browner, with white necklace and unspotted lower breast; bristles are shorter and straighter. White on male's outer tail feathers is restricted to inner webs. (2) Whip-poor-will is smaller, with black throat, U-shaped necklace.

RANGE. Cuba and Hispaniola. **Status**: Common on Cuba, Isla de Pinos, Cayo Coco, and possibly the other large cays north of Ciego de Ávila and Camagüey provinces. Differences in vocalizations and morphology suggest that the Cuban and Hispaniolan populations represent separate species. **Habitat**: Rather dense forest. **Nesting**: Mar–Jul. Nest is on leaf-covered ground. Lays two grayish-green eggs, spotted with brown. **Voice**: A repeated *tu-wurrrr*, often preceded by a quick *tuk* audible only at close range. Heard mainly at dawn and dusk. **Food**: Mainly large insects caught on the wing.

WHIP-POOR-WILL
GUABAIRO CHICO
Caprimulgus vociferus   Pl. 29

DESCRIPTION. 9.5" (24 cm). Mottled dark grayish brown. Throat black with a contrasting white or buff U-shaped necklace. Three outer tail feathers tipped extensively white in male; buffy and more restricted in female. Tail and wings rounded.

SIMILAR SPECIES. (1) Chuck-will's-widow and (2) Greater Antillean Nightjar are larger, with brown throat and bar across base of throat, not U-shaped. Male Chuck-will's-widow has less white on outer tail feathers.

RANGE. Breeds in eastern and southwestern North America, wintering south to Panama. **Status**: Vagrant. One record: 7 Jan (1932). La Habana. **Habitat**: Forests. **Voice**: At night, a repeated *whip-poor-will*. **Food**: Insects.

SWIFTS Apodidae

Small, somewhat swallow-like birds, with tiny bill and legs and short tail. Wings are long, narrow, and pointed, with the wrist close to body. Incredibly fast fliers. Most are dark, with a few white or pale markings. All capture insects on the wing and at least some copulate in flight. Never seen perching on branches or wires, they rest, sleep, and nest on cliffs or in chimneys or hollow trees. Cup-shaped nests are built using salivary secretions and twigs or other plant debris. (**W**:99; **C**:4)

BLACK SWIFT
VENCEJO NEGRO
Cypseloides niger   Pl. 29

DESCRIPTION. 6.5" (17 cm). Almost entirely very dark brown (appears black in flight); whitish forehead, darkening toward crown. Tail forked. Juvenile has white-edged feathers on underparts. Flight less erratic than other swifts.

SIMILAR SPECIES. Male Purple Martin has broader and more triangular wing.

RANGE. Western North America, wintering south to Central America; in Mexico, Central America (Guatemala and Costa Rica), and the Antilles (Cuba, Jamaica, Hispaniola, Puerto Rico, Montserrat, Guadeloupe, Dominica, Martinique, St Lucia and St. Vincent). **Status**: Rare permanent resident in Cuba: Sierra de Guamuhaya, Cienfuegos, and Sancti-Spíritus provinces; and mountains of Holguín, Santiago de Cuba, and Guantánamo provinces. **Habitat**: Over mountain forest. **Nesting**: A nest has not yet been found in Cuba. **Voice**: A low *chip-chip*. **Food**: Insects.

WHITE-COLLARED SWIFT
VENCEJO DE COLLAR
Streptoprocne zonaris   Pl. 29

DESCRIPTION. 8.5" (22 cm). Large and black, with a complete white collar, although not always apparent in young birds. Tail forked. Immature has narrower collar, sometimes reduced to patches or entirely lacking. Very fast fliers, often forming flocks of up to 50.

SIMILAR SPECIES. None.

RANGE. Mexico to northern Argentina; Greater Antilles. **Status**: Rare permanent resident in Cuba: Sierra de Guamuhaya, Cienfuegos, and Sancti-Spíritus provinces; and mountains of Holguín, Santiago de Cuba, and Guantánamo provinces. **Habitat**: Over mountain forest. **Nesting**:

May–Jul. Cliffs near waterfalls, hollow royal palms. Lays two white eggs. **Voice**: A loud, piercing *scree-scree-scree*. **Food**: Insects.

CHIMNEY SWIFT

VENCEJO DE CHIMENEA

Chaetura pelagica   Pl. 29

DESCRIPTION. 5" (13 cm). Dark brown, with paler throat. Tail very short, nearly square with shafts of tail feathers extending beyond the vanes. Sometimes flies in large flocks along the western northern coast, especially during the fall.

SIMILAR SPECIES. None.

RANGE. Central and eastern North America, wintering in Peru, northern Chile, and northwestern Brazil. **Status**: Rare transient in Cuba (10 Oct–Dec; Mar), commonest in coastal areas. **Habitat**: Over cities and towns. **Voice**: A loud, rapid twittering. **Food**: Insects.

ANTILLEAN PALM-SWIFT

VENCEJITO DE PALMA; VENCEJITO

Tachornis phoenicobia   Pl. 29

DESCRIPTION. 4.5" (11 cm). Very small. Dark brown above, with conspicuous white rump and forked tail. Male much darker, with white throat; female is paler with grayish white throat. Both sexes have a dark band across chest. Juvenile similar to female but darker below. Flight erratic, bat like.

SIMILAR SPECIES. None.

RANGE. Cuba, Jamaica, Hispaniola, Saona, and Beata islands, Île-á-Vache. **Status**: Common permanent resident on Cuba and Isla de Pinos. **Habitat**: Over low, flat country such as sugar cane and other agricultural fields, savannas with palms. Also in cities with abundant exotic palms with drooping leaves. **Nesting**: May–Jul. A colonial breeder, building a half cup-shaped nest among dead palm leaves (*Washingtonia*) with plant fibers and feathers. Lays two or three white eggs. **Voice**: Noisy, emitting an almost constant, weak, twittering, *tooee-tooee*. **Food**: Insects.

HUMMINGBIRDS Trochilidae

Very small birds with tiny legs and long, thin bill adapted for sipping nectar from flowers. Stiff, short wings are beaten at very high speed, becoming a blur in flight and producing a characteristic humming sound. Generally colorful, with flashing metallic iridescence. Soft-bodied insects complement their nectar diet. Nests are made and tended exclusively by females; they are simple but beautifully constructed cups, with lichens attached to the outside for concealment. Invariably lays two eggs. Males are usually smaller and more colorful. (**W**:319; **C**:3)

CUBAN EMERALD
ZUNZÚN; ZUMBADOR; PICAFLOR
Chlorostilbon ricordii   Pl. 30

DESCRIPTION. 4″ (10 cm). *Male*: almost entirely dark green with blue iridescence only rarely visible on chest; overall usually appearing black with white undertail coverts. Slim, with long, slightly decurved bill and white spot behind eye, and long, forked, black tail. *Female*: similar, with gray chest and belly. Juvenile similar to female but has duller green on back.
SIMILAR SPECIES. Female Bee Hummingbird is much smaller, with much shorter, white-tipped blue tail.
RANGE. Bahamas and Cuba. **Status**: Common permanent resident on Cuba, Isla de Pinos, and many cays. **Habitat**: Forest, coastal vegetation, gardens. **Nesting**: Year-round. Builds a tiny and deep cup-shaped nest of fine fibers, coated with lichens and spider webs, and placed in a tree or bush 2–4 m above the ground. Lays two white eggs. **Voice**: Male's song is a high-pitched, rapid, rolling series of *slee* notes; also spluttering metallic notes. Female has a destinctive flight call, a high-pitched *seeeee*, repeated two to five times. **Food**: Nectar, insects, spiders.

RUBY-THROATED HUMMINGBIRD

COLIBRÍ

Archilochus colubris Pl. 30

DESCRIPTION. 3.75"(9.5 cm). *Male:* brassy green above; white below. Sides and flanks green; throat brilliant red, contrasting with white border of breast. Tail short and forked; bill thin and slightly decurved. *Female:* similar except tail almost even, and throat and tips of three outer tail feathers white. Immature male resembles female but may have golden cast to upperparts and some red flecks on throat.

SIMILAR SPECIES. Female Bee Hummingbird is smaller, with bluish green back and rounded tail.

RANGE. Southern Canada and eastern United States, wintering to Central America. **Status**: Rare transient to western Cuba (22 Nov; 15 Feb–15 May). **Habitat**: Settled areas. **Voice**: Short buzzy notes. **Food**: Insects, nectar.

BEE HUMMINGBIRD

ZUNZUNCITO; PÁJARO MOSCA; TROVADOR

Mellisuga helenae Pl. 30

DESCRIPTION. 2.5" (6.4 cm). The world's smallest bird. *Male*: iridescent deep blue to green above, gray below. Head, chin, and throat fiery iridescent pink or red. Tail iridescent blue, very short and rounded. Nonbreeding males lack a gorget and so resemble females but have black-tipped tail. Immature male similar to female, with a deeper blue back. *Female*: larger, with bluish green back and gray underparts. Tips of outer tail feathers white.

SIMILAR SPECIES. (1) Female Ruby-throated Hummingbird is larger, with shining green back and slightly forked tail. (2) Female Cuban Emerald is much larger, with iridescent green back and deeply forked black tail.

RANGE. Endemic to Cuba. **Status**: Rare and vulnerable on Cuba and Isla de Pinos. Mainly found today in Guanahacabibes peninsula, Sierra de Anafe, Zapata peninsula, Júcaro, and several eastern mountain ranges, mainly Cuchillas del Toa and Sierra Cristal. **Habitat**: Dense forest, edge of woodlands with plenty of bushes. **Nesting**: Apr–Jun. A neat cup covered with lichens and spider webs. Lays two white eggs. **Voice**: A series of surprisingly loud, very high-pitched, and prolonged whistles and chirps. During the breeding season, males vocalize from the highest leafless branches. Female vocalization similar to male. **Food**: Nectar, insects.

TROGONS Trogonidae

Medium-sized birds, with short, rather broad bills, short wings, and long square tails hanging straight down. Very colorful, but given to perching motionless for long periods. Fruits, flowers, and insects are plucked or captured while fluttering. Sexes are generally different, although this is not the case for Cuba's only species. (**W**:39; **C**:1)

CUBAN TROGON
TOCORORO; TOCOLORO; GUATINÍ

Priotelus temnurus   Pl. 30

Description. 10.5″ (27 cm). This very beautiful species is Cuba's national bird. The iridescent dark green above, with violet blue crown and nape, and white throat and breast, contrasts sharply with vermillion belly. Eye and lower mandible red. Wings and tail intricately patterned in blue, black, green, and white. Tips of tail feathers prominently flared. Flight undulating.
Similar species. None.
Range. Endemic to Cuba. **Status**: Common on Cuba, Isla de Pinos, and some of the larger cays north of Camagüey (Cayo Guajaba and Cayo Sabinal). **Habitat**: Forest of all kinds, including pine. **Nesting**: Apr–Jul. Lays three or four white eggs in abandoned woodpecker holes. **Voice**: A rapid, wooden *to-co-lo* or *to-co-lo-ro*, delivered in series, as well as a hoarse barking and a variety of clucking or chuckling calls. **Food**: Insects, flowers, fruits.

TODIES Todidae

Small, colorful, and compact with short wings and tail, and long straight and flat bills. Todies live in deep forest shade, scanning the foliage for insects. They excavate burrows in rotten wood, or sand or earth banks, and lay two to four white eggs. (**W**:5; **C**:1)

CUBAN TODY
CARTACUBA; PEDORRERA
Todus multicolor   Pl. 30

DESCRIPTION. 4.25" (10,8 cm). Vivid green above; pale gray below with a brilliant red throat patch. Eye blue; sides of neck with blue patch; undertail coverts yellow; sides pink. Juvenile is entirely pale gray below with brown eyes. Perches for long periods, often bobbing head up and down. Flies only short distances, with a peculiar whirring sound produced by the wings.
SIMILAR SPECIES. None.
RANGE. Endemic to Cuba. **Status**: Common on Cuba, Isla de Pinos, and the larger cays north of Camagüey and Ciego de Ávila. **Habitat**: Shady semideciduous forest, coastal vegetation. **Nesting**: Apr–Jul. Lays eggs in burrows excavated into vertical earth banks and rotten logs; also in natural cavities in limestone. On Cayo Coco, successfully excavate in sand, in the entrance of crab burrows. Lays three white eggs. **Voice**: A hard, rapid chatter *tot-tot-tot*. **Food**: Caterpillars, small adult and larval insects, spiders, even small lizards.

KINGFISHERS Alcedinidae

Small to medium-sized, solitary birds with large crested heads, long straight bill, and small legs. Most New World species of kingfishers feed on fish and thus are usually found near water. Sexes usually differ. Kingfishers excavate nesting burrows in banks. (**W**:94; **C**:1)

BELTED KINGFISHER

MARTÍN PESCADOR; PITIRRE DE AGUA

Ceryle alcyon   Pl. 30

DESCRIPTION. 13" (33 cm). Head, back, and breast band bluish gray, with a
wide and nearly complete white collar. Belly entirely white in males, with a
rusty band in females. Juvenile has rusty spotting in the blue breast band.
Perches near water. Wingbeats deep and irregular, showing white wing
patches. Hovers before plunging into water. Prey is devoured on perch.

SIMILAR SPECIES. None.

RANGE. Breeds in North America, wintering sparsely to northern South
America; West Indies. **Status**: Common winter resident and transient in
Cuba, Isla de Pinos, and many cays (year-round). **Habitat**: Seashores with
nearby high vegetation, estuaries, lakes, reservoirs. **Voice**: A resounding
rattle. **Food**: Fish.

WOODPECKERS Picidae

Medium-sized birds with strong pointed bills used for chiseling wood and
bark. Toes are arranged two forward, two backward. Tail feathers are very
stiff, and serve as supports when the bird climbs vertical trunks. Usually
black, gray, and white, with red head markings. Flight is generally short and
undulating. Wood-boring insect larvae are the staple food of most species
and the exceedingly long tongue reaches these through holes in the bark.
Others feed on small insects, sap, or small fruits. Woodpeckers characteristically drum on hollow trunks or other resonant surfaces, and excavate nest
holes in trees. Eggs are white. Sexes usually exhibit at least minor plumage
differences. (**W**:215; **C**:6)

WEST INDIAN WOODPECKER

CARPINTERO JABADO

Melanerpes superciliaris   Pl. 31

DESCRIPTION. 11" (28 cm). Pale gray, with black and white barred back and tail, and red belly. Male has red cap and nape, female has red nape. In flight, outer wing shows white patch. Birds from Isla de Pinos and Cayo Largo are smaller, with less extensive red nape patch. Juvenile similar to adults, but with less red on head.

SIMILAR SPECIES. Yellow-bellied Sapsucker is smaller, with prominent white patch on inner wing and black and white stripes on head.

RANGE. Bahamas, Cuba, and Grand Cayman. **Status**: Common permanent resident on Cuba, Isla de Pinos, and Cayos Coco, Romano, Guajaba, San Felipe, Ávalos, Cantiles, and Largo. **Habitat**: Open forest, palm groves, mangroves. **Nesting**: Feb–Jul. Usually nests in dead palm trees, occasionally very high. Lays five or six white eggs. **Voice**: A loud and frequently repeated *kkrraaa* and other loud, chattering calls, resembling those of the Fernandina's Flicker. **Food**: Usually takes insects and larvae from living branches, trunks, and bromeliads by probing. Also small fruit, lizards, and frogs.

YELLOW-BELLIED SAPSUCKER

CARPINTERO DE PASO; CARPINTERO CHICO

Sphyrapicus varius Pl. 31

DESCRIPTION. 8.5" (22 cm). Back barred black and white; belly pale yellowish white. Head has black and white stripes, and forehead is red. Inner-wing black, with bold white patch especially apparent in flight. Male has red throat. Juvenile is scaled brown and whitish on head, neck, throat, and breast. Wings and tail dull brownish black.

SIMILAR SPECIES. West Indian Woodpecker is larger, with a large red patch on nape and a black patch behind the eye.

RANGE. Breeds in northern North America, wintering south to Central America; West Indies. **Status**: Common winter resident and transient in Cuba, Isla de Pinos, and many cays (3 Oct–1 May). Females and juveniles much more common than adult males. **Habitat**: Widespread in forest, coastal vegetation, and scattered trees. **Voice**: A soft *mew*, somewhat like call of Gray Catbird. **Food**: Pecks horizontal rows of shallow pits in tree bark, sipping the exuded sap and capturing insects attracted to it.

CUBAN GREEN WOODPECKER

CARPINTERO VERDE; GUASUSA; JORRE JORRE; RUÁN

Xiphidiopicus percussus Pl. 31

DESCRIPTION. 9" (23 cm). Vivid olivaceous green back; pale yellow belly, streaked with dark brown. Head mostly white, with a black postocular stripe, and small red patch on base of throat and upper chest. Male has red from forehead to nape. Female has black forecrown streaked with white; back of crown and nape is red. Juvenile duller on back; red on head con-

fined to coronal stripe. Throat and upper chest blackish. The only Cuban woodpecker that produces a wing sound in flight. During the nesting season, the wing noise made by an arriving mate signals the incubating bird to vacate the cavity for incubation changeover. Individuals occurring outside mainland Cuba average somewhat smaller and paler.

SIMILAR SPECIES. None.

RANGE. Endemic to Cuba. **Status**: Common on Cuba, Isla de Pinos, and many cays. **Habitat**: Dry forest, pine and semideciduous woods, coastal vegetation. **Nesting**: Jan–Aug. Nests in both dead and living tree trunks; entrance hole is rather small and round, 2–12 m above ground. Lays three or four white eggs. **Voice**: A loud, repeated *taha-taha*; also a short and harsh, repeated *jorr.* **Food**: Insects extracted from shallow cracks in bark. Also small fruits, frogs.

NORTHERN FLICKER
CARPINTERO ESCAPULARIO
Colaptes auratus   Pl. 31

DESCRIPTION. 13" (33 cm). Beautifully marked, with brown back barred with black, white belly covered with neat black spots. Head mostly brown and gray, with red crescent on nape. A broad black crescent across breast. Male has black mustache-like stripe. When flying, wings and tail are strik-

ingly yellow below. Female lacks mustache. Juvenile paler with plain unmarked head.

SIMILAR SPECIES. None.

RANGE. Most of North America, Central America, Cuba, and Grand Cayman. **Status**: Common permanent resident on Cuba: Sierra del Rosario,
Pinar del Río province; Zapata peninsula; Itabo, Matanzas province;
eastern mountain ranges; Sierra de Trinidad, Sancti-Spíritus, and Cienfuegos provinces; Najasa, Camagüey province; Sierra de Cupeyal; Holguín province; and the larger cays north of Camagüey and Ciego de
Ávila. **Habitat**: Forest, pine woods, coastal vegetation. **Nesting**:
Jan–Aug. Nests in both dead and living trees. Lays four to six white
eggs. **Voice**: An explosive high-pitched *kyeea*. A repeated and increasingly loud *wik-wik-wik-wik-wik*. Also a softer, repeated *wickah, wickah*.
Food: Insects and fruits gathered in trees, or less commonly on the
ground.

FERNANDINA'S FLICKER

CARPINTERO CHURROSO; CARPINTERO DE TIERRA

Colaptes fernandinae   Pl. 31

DESCRIPTION. 13.5" (34 cm). Uniformly yellowish brown barred with
black; cinnamon head. Male has black mustache-like stripe, lacking in females. Yellowish on undersurface of wings and tail, as in Northern Flicker.
Juvenile is duller.

SIMILAR SPECIES. None.

RANGE. Endemic to Cuba. **Status**: Rare and vulnerable: Sierra del Rosario,
Pinar del Río province; Nueva Paz, La Habana province; Zapata peninsula,
Matanzas province; El Dorado and Corralillo, Villa Clara province; Najasa,
Camagüey province; and northern Holguín province. **Habitat**: Rather
open forest with abundant palms; edges of savannas. **Nesting**: Feb–May.
Nests in dead palms, entrance hole quite large. Lays four or five white
eggs. **Voice**: Similar to Northern Flicker. A repeated and increasingly loud
wik-wik-wik-wik-wik. Also a loud *keer*, and repeated chattering and chur-

ring calls, *krrr*, resembling those of the West Indian Woodpecker. **Food**: Mostly ants and insect larvae. Usually feeds on the ground.

IVORY-BILLED WOODPECKER

CARPINTERO REAL

Campephilus principalis Pl. 31

Description. 19″ (48 cm). Black with white stripes along neck and back; a broad white stripe on wings. Tall crest is red in male, black in female. Bill heavy, straight, and ivory colored.

Similar species. None.

Range. Formerly southeastern United States, and Cuba. Now officially considered extirpated from United States. **Status**: Critically endangered. Last century reported from Pan de Guajaibón, Ensenada de Cochinos, and Guantánamo mountains. A few individuals may survive in Cuchillas del Toa mountains where at least one male and one female were observed in 1987 in a pine forest near an abandoned mining camp named Los Rusos on the border between Holguín and Guantánamo provinces. Unsuccessful expeditions in search of this species were conducted in 1991, 1992, and 1993, but in 1998 new evidence indicating the probable presence of this species was found in the Sierra Maestra. **Habitat**: Tall, undisturbed semideciduous and pine forests, with an abundance of dead trees. **Nesting**: Apr–Jun. Nests in tall, old pines. Lays two or four white eggs. **Voice**: A very distinctive, repeated, high-pitched, nasal, *pent*. **Food**: Large insect larvae extracted from rotting logs.

TYRANT FLYCATCHERS Tyrannidae

Small to medium-sized birds, with few salient structural characters. Most are colored in greens, grays, black, pale yellows, and white. Bill is broad at the base, flat, and slightly hooked at the tip, and in most species is surrounded by prominent rictal bristles. Some flycatchers have a short crest and a small brightly colored crown patch, visible only when seen from above at close range. They perch on exposed branches for long periods, darting out to catch flying insects, often returning to the same perch. Solitary and territorial. Sexes almost invariably alike. (**W**:537; **C**:18)

WESTERN WOOD-PEWEE
BOBITO DE BOSQUE DEL OESTE

Contopus sordidulus Pl. 32

DESCRIPTION. 6.25″ (16 cm). Almost identical to Eastern Wood-Pewee. Safely distinguished only by voice. Brownish green above, with whitish throat and belly. Wing bars distinct. Chest usually has a broad and uninterrupted dark-gray band. Bill generally entirely black. Juvenile has cinnamon-edged feathers on back and wings.

SIMILAR SPECIES. (1) On average, Eastern Wood-Pewee has narrower chest band and yellower lower mandible. (2) Cuban Pewee has bright whitish crescent behind eye and a broader, flatter bill.

RANGE. Western North America and Central America, wintering from northern South America to Bolivia. **Status**: Rare transient in Cuba and some cays (4 Jul–22 Oct; 31 Mar–22 Apr). **Habitat**: Open forests, cities. **Voice**: Most common call is a hoarse descending *pheeer*. **Food**: Mainly insects caught on the wing.

EASTERN WOOD-PEWEE
BOBITO DE BOSQUE

Contopus virens Pl. 32

DESCRIPTION. 6.25″ (16 cm). Almost identical to Western Wood-Pewee. Safely distinguished only by voice. Olivaceous brown above, with whitish throat and belly. The broad, dark gray chest band is usually interrupted centrally. Lower mandible mostly yellow. Wing bars whitish. Juvenile as Western Wood-Pewee.

SIMILAR SPECIES. (1) On average, Western Wood-Pewee has broader chest band, and blacker lower mandible. (2) Cuban Pewee is paler, and has a bright whitish crescent behind eye and a broader, flatter bill.

RANGE. Breeds in eastern North America, wintering in South America from Colombia and Venezuela to Peru and western Brazil. **Status**: Common transient in Cuba, Isla de Pinos, and some cays (24 Aug–4 Nov; 26 Mar–22 Apr). One sighting 29 Dec, in Zapata peninsula. **Habitat**: Open forests, gardens. **Voice**: A clear, plaintive, slurred *pee-a-wee* (rising); also a descending *pee-ur*. **Food**: Mainly insects caught on the wing.

CUBAN PEWEE
BOBITO CHICO; PITIBOBO

Contopus caribaeus Pl. 32

DESCRIPTION. 6.25″ (16 cm). Dark olivaceous gray above, with rather dark grayish-buff chest and belly. A whitish crescent immediately behind eye.

Bill broad and flat; lower mandible yellow. Quivers tail upon alighting. Juvenile has paler lower mandible and whitish buff wing bars.

SIMILAR SPECIES. (1) La Sagra's Flycatcher is larger, with rufous on wing and tail feathers. Head and bill are wholly dark. Call notes are similar, but are delivered more slowly. (2) Both wood-pewees have narrower bill, dark gray chest, and lack mark behind the eye. (3) Eastern Phoebe is paler below with narrower black bill and unmarked eye.

RANGE. Bahamas and Cuba. **Status**: Common permanent resident on Cuba, Isla de Pinos, and several cays. **Habitat**: Semideciduous woods, pine forests, forest edge, swamps, mangroves. Rare at high elevations. **Nesting**: Mar–Jun. Nest is cup shaped, on a limb of a tree or in fork of tree or shrub. Built with fine rootlets and hairs, covered with lichen and moss. Lays up to four white eggs, heavily spotted with dark, violet, and brown at larger end. **Voice**: A prolonged and descending whistle, *weeeeooooo*. Also a thin, repeated *weet*, on one pitch, similar to, but faster than that of La Sagra's Flycatcher. **Food**: Mainly insects caught on the wing.

YELLOW-BELLIED FLYCATCHER

BOBITO AMARILLO

Empidonax flaviventris   Pl. 32

DESCRIPTION. 5.5" (14 cm). Greenish olive above; yellowish below, with a broad olive wash on chest. A conspicuous yellow eye ring. Lower mandible pale orange. Two well-marked yellowish or whitish wing bars. Legs gray.

SIMILAR SPECIES. (1) Acadian Flycatcher is paler yellow below, with whitish throat, white lower breast, and longer wings. (2) Alder and (3) Willow Flycatchers usually lack distinct eye ring and have whitish throat, yellowish belly, and brownish olive above. All three species lack Yellow-bellied Flycatcher's overall decidedly yellow cast.

RANGE. Breeds in northern North America, wintering in Middle America. **Status**: Very rare fall transient in Cuba (8 Sep–4 Oct). **Habitat**: Open forests, thickets, gardens, coastal vegetation. **Voice**: A whistled *chu-wee*, and a resonant *che-lek*. **Food**: Mainly insects caught on the wing.

ACADIAN FLYCATCHER

BOBITO VERDE

Empidonax virescens Pl. 32

DESCRIPTION. 6" (15 cm). Olivaceous green above, with whitish throat. Chest with olive wash; upper belly white; belly and undertail coverts yellowish. Legs gray. Lower mandible yellow. Yellow eye ring. Two conspicuous wing bars. Immature has strong wash of yellow below. Juvenile has brownish olive upperparts, and is whiter below, with buffy wing bars.

SIMILAR SPECIES. (1) Yellow-bellied Flycatcher is much yellower below, with shorter wings and a more conspicuous eye ring. (2) Alder and (3) Willow Flycatchers are darker and browner above, with eye ring indistinct or lacking.

RANGE. Southern Ontario, eastern United States, wintering from Nicaragua to northern South America. **Status**: Regular but rare transient in Cuba, mostly in fall (6 Sep–15 Oct; 28 Apr). **Habitat**: Open forests, gardens. **Voice**: A rapid ascending whistle *pee-peet-sa*. **Food**: Mainly insects caught on the wing.

ALDER FLYCATCHER

BOBITO DE ALDER

Empidonax alnorum Pl. 32

DESCRIPTION. 5.75" (15 cm). Virtually identical to Willow Flycatcher. Safely differentiated only by voice. Brownish olive above. Eye ring indistinct or lacking. Wing bars whitish and conspicuous. Whitish throat and pale olive breast; pale yellow belly. Legs black, lower mandible yellow. Juvenile has buff wing bars and indistinct eye ring.

SIMILAR SPECIES. (1) On average, Willow Flycatcher has very inconspicuous or absent eye ring. (2) Acadian Flycatcher and (3) Yellow-bellied Flycatcher are olive above, with a very distinct eye ring. (4) Wood-pewees are larger with slightly crested heads, longer wings, and gray chest bands.

RANGE. Breeds in Alaska, Canada, and northeastern United States, wintering in South America. **Status**: Very rare transient in Cuba (Oct). **Habitat**: Open forests, gardens. **Voice**: A buzzy *ree-bee-o*; the call, a loud *peep*. **Food**: Mainly insects caught on the wing.

WILLOW FLYCATCHER

BOBITO DE TRAILL

Empidonax traillii Pl. 32

DESCRIPTION. 6" (15 cm). Virtually identical to Alder Flycatcher. Safely differentiated only by voice. Brownish olive above. Eye usually without distinct ring. Two conspicuous wing bars. Throat and upper belly white. Chest has an olivaceous wash, belly a yellow wash. Legs black; lower mandible yellow.

SIMILAR SPECIES. (1) On average, Alder Flycatcher has more distinct eye ring. (2) Acadian and (3) Yellow-bellied Flycatchers are greener above with a conspicuous yellowish eye ring.

RANGE. Western Canada, northern and central United States, wintering south to Central America. **Status**: Very rare fall transient in Cuba (12 Sep–15 Oct). **Habitat**: Open forests, gardens. **Voice**: a buzzy *phitz-bew*, call an emphatic *rit* or *whit*. **Food**: Mainly insects caught on the wing.

EASTERN PHOEBE

BOBITO AMERICANO

Sayornis phoebe Pl. 32

DESCRIPTION. 7" (18 cm). Grayish brown above, with dark head, lacking eye ring. Throat white and belly pale yellowish gray. Two inconspicuous wing bars; black bill. Juvenile is browner above, with two buff wing bars. Tail-wagging habit is a good field mark.

SIMILAR SPECIES. (1) Cuban Pewee has distinct white crescent immediately behind eye, and broad flat bill with yellow lower mandible. (2) Wood-pewees have wing bars, are darker on chest, and do not wag their tails.

RANGE. Canada, eastern and central North America, wintering south to Oaxaca, Mexico. **Status**: Vagrant. Five records: Feb (last century); 14, 16 Sep (1960); 28 Nov; one undated. La Habana and Sierra de Najasa, Camagüey. **Habitat**: Open forests. **Voice**: A repeated, raspy *phoe-be*. **Food**: Mainly insects caught on the wing.

GREAT CRESTED FLYCATCHER

BOBITO DE CRESTA

Myiarchus crinitus Pl. 32

DESCRIPTION. 8" (20 cm). Olive head and back; gray from chin to chest; rather bright yellow belly, Outer webs of primaries rufous; all but central tail feathers have rufous inner webs. Bill black with orange base to lower mandible.

SIMILAR SPECIES. La Sagra's Flycatcher has paler gray chest, paler yellow belly, and dark cinnamon brown inner webs of wings and tail feathers.

RANGE. Central and eastern North America, wintering to northern South America. **Status**: Very rare transient in Cuba (20 Sep–10 Nov; Mar–22 Apr). **Habitat**: Thick forests, often seen in tops of trees. **Voice**: Varied: a loud whistle, *wheep*; a rolling *prrr-eet*; and a *wheerrup*. **Food**: Mainly insects.

LA SAGRA'S FLYCATCHER

BOBITO GRANDE

Myiarchus sagrae Pl. 32

DESCRIPTION. 8" (20 cm). Brownish olive back; gray from chin to upper belly; very pale yellow lower belly. Wings and tail have dark cinnamon brown inner webs. Bill entirely black.

SIMILAR SPECIES. (1) Great Crested Flycatcher has darker gray chest, brighter yellow belly, and bright rufous in wings and tail. (2) Cuban Solitaire has creamy white eye ring and whisker marks; commonly perches with hanging tail. (3) Cuban Pewee is smaller, with a white crescent behind the eye, and yellowish lower mandible.

RANGE. Bahamas, Cuba, and Grand Cayman. **Status**: Common permanent resident on Cuba, Isla de Pinos, and many of the larger cays. **Habitat**: Forests, from sea level to rather high elevations. **Nesting**: Mar–Jul. In dead branch holes, woodpecker holes, the bottom is lined with dry grasses, hairs, rootlets, and feathers. Lays four yellowish-white eggs, spotted with brown and violet at larger end. **Voice**: Song, a whistled *weeet-ze-weer* or *weeet-ze*. Call, a slightly ascending whistled *weet*, repeated at intervals, sometimes with a squealing tone. Compare with Cuban Pewee. **Food**: Insects, either catching them on the wing or gleaning them from leaves or twigs; also small lizards.

TROPICAL KINGBIRD
PITIRRE TROPICAL
Tyrannus melancholicus Pl. 33

DESCRIPTION. 9.25" (23 cm). Grayish green above, with gray head. Bill heavy; throat white. Breast yellowish olive; belly bright yellow. Tail notched, brownish black. Concealed crown patch red. Juvenile similar to adult, but without crown patch and with buff-edged wing coverts.

SIMILAR SPECIES. Western Kingbird has white-edged, squared black tail.

RANGE. Arizona and northwestern Mexico to Argentina. **Status**: Vagrant. Three records: 17 Feb; two undated. Zapata peninsula and Cayo Largo. **Habitat**: Open areas. **Voice**: Twittering, repeated *pip*. **Food**: Mainly insects caught on the wing.

WESTERN KINGBIRD*
PITIRRE DEL OESTE
Tyrannus verticalis Pl. 33

DESCRIPTION. 8.5" (22 cm). Olivaceous above with pale gray head and yellow belly. Bill moderately proportioned. Tail squared, black with narrow white margins. Crown patch orange, usually concealed. Juvenile similar to adult but without crown patch and with buff-edged wing coverts.

SIMILAR SPECIES. Tropical Kingbird has notched and entirely brownish black tail, olive wash across breast, and considerably larger bill.

RANGE. Breeds in western North America, wintering south to Costa Rica. **Status**: One sight record for Cuba: 26 Aug (1986). Najasa. **Habitat**: Open country. **Voice**: A short emphatic *whit*. **Food**: Mainly insects caught on the wing.

EASTERN KINGBIRD

PITIRRE AMERICANO

Tyrannus tyrannus   Pl. 33

DESCRIPTION. 8.5" (22 cm). Dark slaty gray above; white below, with a pale gray wash on chest. Head black, with concealed orange crown patch. Wing bars indistinct. A very distinct white terminal band on square tail. Immature is slightly brownish gray above and lacks crown patch.

SIMILAR SPECIES. (1) Loggerhead Kingbird has buff white terminal band on tail, a yellow wash on axillaries and undertail coverts, and a much larger bill. (2) Gray Kingbird has gray head and back, notched tail without white, and much larger bill.

RANGE. North America, wintering south to northern Chile and northern Argentina. **Status**: Rare transient in Cuba; somewhat more frequent in western regions (4 Jul–17 Oct; 31 Mar–22 Apr). **Habitat**: Open forests, farmland, parks, towns. **Voice**: A strident shrill, *kip-kip-kipper*, or *dzee-dzee*. **Food**: Insects caught on the wing, fruit.

GRAY KINGBIRD

PITIRRE ABEJERO; PITIRRE

Tyrannus dominicensis   Pl. 33

DESCRIPTION. 9" (23 cm). Dark gray above, with a black ear patch. Underparts mostly white, with a faint yellow wash on belly and undertail coverts. Tail notched. Bill large. Concealed orange crown patch. Juvenile has brown-edged feathers on wings and back and lacks crown patch.

SIMILAR SPECIES. (1) Eastern and (2) Loggerhead Kingbirds have black head and squared tail. (3) Giant Kingbird is larger with larger bill, black head, and more square-ended tail.

RANGE. Breeds in coastal southeastern United States, West Indies, northern South America, wintering from Hispaniola and Puerto Rico south through the Lesser Antilles and from Panama to northern South America. **Status**: Common summer resident and transient in Cuba, Isla de Pinos, and many larger cays (19 Feb–6 Oct). **Habitat**: Open forests and farmland, parks, towns. **Nesting**: Apr–Jul. Builds cup-shaped nest in trees, with twigs, vines and grasses; lined with fine grass and rootlets. Lays three or four creamy white or pale pink eggs, speckled and spotted with chestnut red, pale lilac, and gray. **Voice**: A loud and rolling *pit-piteerri-ri-ree* higher pitched than Loggerhead Kingbird; also a short, fast *pi-tir-re*. **Food**: Bees, other flying insects. Also lizards, small fruits.

LOGGERHEAD KINGBIRD

PITIRRE GUATÍBERE; PITIRRE CANTOR; GUATÍBERE

Tyrannus caudifasciatus   Pl. 33

DESCRIPTION. 9″ (23 cm). Dark gray tail and back contrast with black head. White below, with a yellow wash on axillaries and undertail coverts. Concealed yellow or pale-orange crown patch. Tail squared, with buff white terminal band. Juvenile has brownish wing bars and lacks crown patch.

SIMILAR SPECIES. (1) Gray Kingbird has gray head and a prominent black auricular patch. (2) Giant Kingbird is larger, with heavier bill and less conspicuous whitish-tipped tail. Head coloration contrasts only weakly with back. (3) Eastern Kingbird is whiter below, with bolder, whiter tail band and considerably smaller bill.

RANGE. Bahamas, Greater Antilles, and Cayman Islands. **Status**: Common permanent resident in Cuba, Isla de Pinos, and several large cays in Sabana-Camagüey and Jardines de la Reina archipelagos. **Habitat**: Forests, mangroves, swamp edges. **Nesting**: Apr–Jul. A cup-shaped nest is usually built high above the ground, with twigs, rootlets, and hairs without lining. Lays two or three salmon-colored eggs with reddish brown and violet markings at the broad end. **Voice**: A loud, rising, *pit-pit-pit-pit-pit-tirr-ri-ri-reee* and a rolling, almost spluttering chatter. **Food**: Large flying insects, lizards, small fruits.

GIANT KINGBIRD
PITIRRE REAL

Tyrannus cubensis   Pl. 33

DESCRIPTION. 10.25 ″ (26 cm). A massive kingbird with a very heavy bill. Dark gray above with black head and white underparts. Tail slightly notched, sometimes with pale tip. Concealed orange crown patch, lacking in juvenile.

SIMILAR SPECIES. (1) Loggerhead Kingbird is smaller, with darker head, shorter bill, and a somewhat more distinct white terminal band on squared tail. (2) Gray Kingbird is smaller, with gray head and well-notched tail.

RANGE. Cuba. Formerly Great Inagua and Caicos Islands. **Status**: Very rare and endangered permanent resident in Cuba and Isla de Pinos: Guanaha-cabibes peninsula; Sierra de Anafe; Sierra de Najasa; Cabo Cruz; mountains of the eastern provinces (Ojito de Agua); and Santa Fé River, Isla de Pinos. **Habitat**: Dry savanna with scattered *Ceiba* trees; tall forests near rivers; also pine forests. **Nesting**: Mar–Jun. Builds cup-shaped nest high in trees, made of twigs, rootlets, and grasses without lining. Lays two or three creamy-white eggs speckled and spotted with chestnut red, pale lilac, and gray. **Voice**: A loud, burry *tooe-tooe-tooee-tooee-tooee*. **Food**: Large insects, lizards, small nestling birds.

SCISSOR-TAILED FLYCATCHER
BOBITO COLA DE TIJERA

Tyrannus forficatus   Pl. 32

DESCRIPTION. 13 ″ (33 cm, including long tail feathers of male). Pale gray; wings, and very long forked tail, black. Lower belly and undertail coverts washed with pink and underwing coverts also pink, with red axillary feathers. Female has shorter tail and immature has still shorter tail.

SIMILAR SPECIES. Fork-tailed Flycatcher has black cap and lacks pink tones.

RANGE. Texas and adjacent parts of the United States and northeastern Mexico, wintering south to Panama. **Status**: Vagrant. Four records: 21 Nov (1952); 11 Nov (1984); two undated. Ensenada de Guadiana and San Antonio de los Baños. **Habitat**: Open or semiopen country. **Voice**: A harsh *kee-kee*. **Food**: Mainly insects caught on the wing.

FORK-TAILED FLYCATCHER
BOBITO DE COLA AHORQUILLADA
Tyrannus savana   Pl. 32

DESCRIPTION. 14" (36 cm, including long tail feathers). Black above, except pale gray back; white below. Tail extremely long and forked. Immature has shorter tail.

SIMILAR SPECIES. Scissor-tailed Flycatcher has white head and pink sides.

RANGE. Central and South America, wintering irregularly through Middle America from breeding areas in southeastern Mexico south to central Panama. In South America from Colombia to Argentina. **Status**: Vagrant. Two records: 11 Nov; Feb. Western provinces. **Habitat**: Open country. **Food**: Mainly insects caught on the wing.

VIREOS Vireonidae

Small birds, with rather heavy, slightly hooked bills. Inconspicuously colored, generally brownish or grayish green above and pale below. Species with eyebrows lack wing bars, while those with spectacles or eye rings have wing bars. Food is principally insects, especially insect larvae (caterpillars) gleaned from foliage, and occasionally small fruits. Nests are cup-shaped and rather large. Vireos superficially resemble warblers, differing in their heavier bills and more deliberate movements. Sexes alike. (**W**:51; **C**:9)

WHITE-EYED VIREO
VIREO DE OJO BLANCO
Vireo griseus   Pl. 34

DESCRIPTION. 5.25" (13 cm). Greenish gray above; white below, with yellow sides. White eye surrounded by yellow spectacles. Two white wing bars. Immature has brown eye.

SIMILAR SPECIES. (1) Thick-billed Vireo is yellowish or whitish below, with heavier bill, more distinct wing bars, and brown eye. (2) Solitary Vireo has white spectacles on gray face and bold wing bars.

RANGE. Breeds in eastern North America and Mexico, wintering south to Nicaragua, Bermuda, Bahamas, Cuba, Grand Cayman, and Swan Islands. **Status**: Common winter resident on Cuba, Isla de Pinos, and some larger cays (16 Sep–22 Apr). **Habitat**: Thickets, shrubbery. **Voice**: A distinct and rapid *chick-a-per-weeoo-chick*. **Food**: Insects, fruits.

THICK-BILLED VIREO
VIREO DE LAS BAHAMAS
Vireo crassirostris Pl. 34

DESCRIPTION. 5.5ʺ (14 cm). Olive above; pale yellow or whitish below, occasionally with olive wash to sides and flanks. Head gray. Lores black. Spectacles are boldly yellow between the eye and bill, extending to yellow around eye, or merely to white crescents in some individuals. Two distinct yellowish-white wing bars; brown eye; thick bill. Juvenile resembles adult but head and back feathers tipped brown, and wing bars poorly defined.

SIMILAR SPECIES. (1) Immature White-eyed Vireo has yellow on sides only, smaller head and bill, and less distinct wing bars. (2) Yellow-throated Vireo has yellow breast and white belly; olive head contrasts strongly with gray back and rump.

RANGE. Bahamas, Cuba, Cayman Islands, Tortue Island (off Hispaniola), and on Providencia and Santa Catalina islands in the western Caribbean Sea. **Status**: Common permanent resident on Cayo Paredón Grande. Also records of immatures on Cayo Coco. Critically endangered. **Habitat**: Coastal vegetation. **Nesting**: Mar–May. Nest and eggs undescribed in Cuba. **Voice**: Extremely variable. A sharp *chi-chip-weeeo-chip*. Calls include a buzzy *shhh*, a low *turrrr*, and a nasal *enk*, very similar to that of the Red-breasted Nuthatch (*Sitta canadensis*). **Food**: Insects.

CUBAN VIREO
JUAN CHIVÍ; OJÓN; CHICHINGUAO
Vireo gundlachii Pl. 34

DESCRIPTION. 5.25ʺ (13 cm). Dark olive gray above; pale yellow below. Lores and large postocular crescent creamy white. Eye is pale to reddish brown and appears proportionately large in comparison to other vireos

due to narrow area of brown, bare skin around the eye. One or two faint wing bars. Individuals on Isla de Pinos may be whitish below. Juvenile is duller.

SIMILAR SPECIES. Only vireo in Cuba with combination of large eye, creamy lores and postocular crescent, and faint wing bars.

RANGE. Endemic to Cuba. **Status**: Common on Cuba; also present in Isla de Pinos and several larger cays. **Habitat**: Forest, thickets, bushes, mainly at low elevations. **Nesting**: Mar–Aug. Nest is cup shaped, made of grasses, moss, lichen, and hair. Lays three white, brown-spotted eggs. **Voice**: Song, a loud whistling *chuee-chuee*, or *see-see-ri-lo*, or *wee-beee-eeer*, or *wheee-tzeeooo*; highly variable. Calls include a rapid, and slightly descending series of *chi* notes similar to Yellow-throated Vireo. Also a scolding and often repeated *kik* note, a soft rattling call, and occasionally a metallic, repeated *poing*. During spring courtship, rapid, repeated *wheet* notes, also a muted *pewt* repeated at intervals. **Food**: Insects, fruits, small lizards.

BLUE-HEADED VIREO

VERDÓN DE CABEZA GRIS

Vireo solitarius   Pl. 34

DESCRIPTION. 5.5″ (14 cm). Olivaceous above; head bluish gray with bold and contrasting white spectacles. Throat and central underparts pure white; sides yellow. Two distinct yellow wing bars.

SIMILAR SPECIES. (1) White-eyed Vireo has white eye and yellow spectacles. (2) Thick-billed Vireo has portion of spectacle in front of eye strongly yellow. (3) Yellow-throated Vireo has yellow spectacles and yellow throat and breast.

RANGE. Breeds in North America and Central America south to Honduras, wintering south to Costa Rica, possibly western Panama. **Status**: Very rare winter resident in Cuba and Isla de Pinos (18 Nov–19 Apr). **Habitat**: Forest, dense thickets and bushes. **Voice**: Scold note is a burry, descending *jeeeer-jeer-jeer.* **Food**: Insects, fruits.

YELLOW-THROATED VIREO
VERDÓN DE PECHO AMARILLO

Vireo flavifrons Pl. 34

DESCRIPTION. 5.5" (14 cm). Olive above with gray rump and very bright yellow throat, breast, and spectacles. White belly and wing bars. Immature is duller.

SIMILAR SPECIES. (1) Thick-billed Vireo is entirely pale yellow below with gray head and more distinct black lores. (2) Pine Warbler has thinner bill, green rump, and streaked sides.

RANGE. Breeds in eastern North America, wintering to northern South America; Bahamas. **Status:** Rare winter resident in Cuba, Isla de Pinos, and some larger cays. Common in Cayo Coco (3 Aug–8 May). **Habitat:** Open semideciduous forest. **Voice:** Song, often heard in winter in Cuba, is burry pairs of notes with a distinct pause between the notes *zeeeooo-*(pause)-*zeeear*, with a slight rise in inflection on the second note. Call is a descending, series of *chi* notes, slower and more strongly descending than that of Cuban Vireo. **Food:** Insects, fruits.

WARBLING VIREO
VIREO CANTOR

Vireo gilvus Pl. 34

DESCRIPTION. 5.5" (14 cm). Olivaceous gray above with gray crown; white below, with pale yellow sides. Eyebrow whitish, not black bordered above, and not contrasting strongly with rest of head, giving the face a blank look. Lores and wings unmarked.

SIMILAR SPECIES. (1) Philadelphia Vireo has shorter bill, dark lores, and brighter and more extensively yellow underparts. (2) Red-eyed and (3) Black-whiskered Vireos have distinct white eyebrow bordered above and below with narrow black lines.

RANGE. Breeds in North America and Mexico, wintering south to Nicaragua. **Status:** Very rare transient in Cuba (1 Sep–26 Oct). **Habitat:** Forests, gardens. **Voice:** A complaining *shway* similar to notes of Philadelphia, Red-eyed, and Black-whiskered Vireos. **Food:** Insects, fruits.

PHILADELPHIA VIREO
VIREO DE FILADELFIA

Vireo philadelphicus Pl. 34

DESCRIPTION. 5.25" (13 cm). Closely resembles Warbling Vireo but slightly smaller with shorter bill. Olivaceous above without wing bars; more or less strongly and uniformly yellow below. A gray cap, with white eyebrow and dark eyeline continuing through lores.

SIMILAR SPECIES. (1) Warbling Vireo has unmarked pale-gray lores and mostly white underparts, with yellow wash confined to sides. Cap contrasts less with olivaceous gray back. (2) Winter Tennessee Warbler usually has brighter green upperparts, paler underparts, thinner bill, and is less deliberate in its movements.

RANGE. Breeds in northern North America, wintering from Guatemala to central Panama. **Status**: Rare transient in Cuba; one winter record (6 Oct–1 Nov; 18 Feb). **Habitat**: Open areas with some trees, gardens, forests. **Voice**: A *shway* note similar to Warbling, Red-eyed, and Black-whiskered Vireos. **Food**: Insects, fruits.

RED-EYED VIREO

VIREO DE OJO ROJO

Vireo olivaceus   Pl. 34

DESCRIPTION. 6" (15 cm). Olivaceous green above; white below; some with yellowish sides. Gray cap and distinct white eyebrow outlined in black. Eye red. Dwells high on treetops where it moves deliberately through the foliage. Immature has brown eye and pale yellow flanks and undertail coverts.

SIMILAR SPECIES. (1) Black-whiskered Vireo has whisker stripe and longer bill, and the black line above the eyebrow is less marked. (2) Warbling Vireo has diffuse white eyebrow without black margins.

RANGE. Breeds in North, Central, and South America; North American populations winter in South America, south to Brazil. **Status**: Common transient in Cuba, Isla de Pinos, and some larger cays (28 Aug–13 Nov; 14 Feb–20 Apr). **Habitat**: Forests, gardens. **Voice**: The classic *shway* of this group of vireos. **Food**: Insects, fruits.

BLACK-WHISKERED VIREO

BIEN-TE-VEO; PREDICADOR; CHICHINGUAO

Vireo altiloquus   Pl. 34

DESCRIPTION. 6.5" (17 cm). Olivaceous green above; whitish below, with yellowish sides. Gray cap and white eyebrow bordered by two narrow black lines; narrow black whisker. Eye red. Immature has brown eye. Juvenile lacks black whisker.

SIMILAR SPECIES. (1) Red-eyed Vireo has shorter bill and lacks whisker stripe, and its eyebrow is outlined more conspicuously in black. (2) Warbling Vireo is smaller and plainer with olivaceous gray head and poorly defined white eyebrow.

RANGE. Breeds in southern Florida and West Indies; winters in South America south to eastern Brazil, rarely on Hispaniola, Puerto Rico, and northern Lesser Antilles. **Status**: Common summer resident in Cuba, Isla de Pinos, and many of the larger cays (22 Feb–21 Oct). Commonly arrives on or around 10 Mar. Earliest arrivals are in western mountain regions. **Habitat**: Low coastal vegetation, semideciduous woods, gardens, mangroves. **Nesting**: Apr–Jul. Nest is cup shaped. Lays two or three white eggs with black and purplish brown speckles. **Voice**: Frequently repeated two-, three-, or four-syllable phrases *tsee-tsee-we* louder and more repetitious and more warbling than the song of Red-eyed Vireo. **Food**: Insects, fruits.

CROWS Corvidae

Large birds, with heavy, pointed bills and powerful legs. Wings and tail are broad and rounded. Crows are gregarious and noisy as a rule, and many species are entirely black. Omnivorous, they eat fruits, large insects, small vertebrates as well as other birds' eggs and young. Nests are crudely assembled, and usually concealed in high and dense foliage. Sexes are alike. (**W**:647; **C**:2)

PALM CROW
CAO PINALERO; CAO RONCO
Corvus palmarum   Pl. 35

DESCRIPTION. 17" (43 cm). Entirely black, with a very slight metallic sheen. Juvenile has a somewhat duller tone. Small bill; short wings. May form small flocks. Often seen on the ground. Typically flicks tail upward on alighting.

SIMILAR SPECIES. Cuban Crow is slightly larger with wings extending nearly to tip of tail and with longer bill. Feathers covering base of upper mandible are shorter, not covering the nostrils. Rarely seen on ground. Often best differentiated by voice.

RANGE. Cuba and Hispaniola. **Status**: Vulnerable. Restricted to the vicinity of Mina Dora in Pinar del Río province, and Sierra de Najasa, Tayabito, Miguel, and El Jardín in central Camagüey province. The Cuban and Hispaniolan populations are vocally distinct and some authors believe that the two taxa are best considered as species. **Habitat**: Forests near pine plantations; farms, parks, towns. **Nesting**: Mar–Jul. On the fronds of royal palms, and very similar to that of Cuban Crow. Eggs in Cuba not yet described. **Voice**: A nasal, complaining *craaah* note, sometimes repeated in pairs, vaguely reminiscent of the voice of Fish Crow (*Corvus ossifragus*) in North America. **Food**: Omnivorous.

CUBAN CROW
CAO MONTERO; CUERVO
Corvus nasicus   Pl. 35

DESCRIPTION. 18″ (46 cm). Entirely black, with a slight metallic sheen. Juvenile is duller. Usually occurs singly or in pairs, but larger groups may gather, especially near communal roosts. Noisy. Rarely seen on ground.

SIMILAR SPECIES. Palm Crow is slightly smaller, with shorter wings and bill. Feathers covering base of upper mandible are longer, covering the nostrils. Regularly seen on ground. Often best differentiated by voice. Palm Crow is much less widely distributed in Cuba, being restricted to two small mountain regions.

RANGE. Cuba and Caicos Islands. **Status**: Common resident on Cuba, Isla de Pinos, and some larger cays north of Camagüey and Ciego de Ávila. Especially common in Zapata peninsula, Sierra de Najasa, and Sagua-Baracoa. **Habitat**: Forests with scattered clearings, palm groves, borders of swamps. **Nesting**: Mar–Jul. Builds nest on large bromeliads or among palm fronds, made of twigs, dry grasses, and feathers. Lays up to four greenish eggs, spotted with brown and lilac. **Voice**: Very noisy, producing a variety of rather high-pitched, warbled, and often parrot-like calls, sometimes rhythmically repeated, *kweaa*, *kraak*, or *kaaaa*. Common call is *gawwaaak-gawow*. **Food**: Omnivorous.

SWALLOWS Hirundinidae

Small to medium-sized birds, with long, pointed wings. Bills are small, with wide gape. Legs are short. Most species are dark above and white or off-white below. Food is mainly insects, captured in flight. Gregarious and generally migratory, swallows frequently perch on wires or in leafless trees. Some nest in tree holes; others build more-or-less spherical mud structures. In most species, sexes are similar. (**W**:93; **C**:9)

PURPLE MARTIN
GOLONDRINA AZUL
Progne subis   Pl. 36

DESCRIPTION. 8″ (20 cm). *Male*: entirely very dark metallic purplish blue. *Female*: very dark iridescent purplish blue above; throat, grayish brown breast, and sides; fading to white on streaked belly. Tail forked. Juvenile similar to female. In flight, fast flapping alternates with short glides.

SIMILAR SPECIES. (1) Male Cuban Martin is virtually indistinguishable, but female lacks streaks on belly. (2) Black Swift is dark brown, with narrower and longer wings and faster flight.

RANGE. Breeds in North America, wintering from Colombia and Venezuela south to Brazil. **Status**: Common transient in Cuba, Isla de Pinos, and some cays (12 Aug–6 Nov; 9 Feb–Mar). **Habitat**: Savannas, swamps, rice fields, suburban and rural environments. **Voice**: A rich repeated *tew* and melodious warbled notes with a throaty, plucking quality. **Food**: Insects caught on the wing.

CUBAN MARTIN
GOLONDRINA AZUL CUBANA
Progne cryptoleuca   Pl. 36

DESCRIPTION. 8″ (20 cm). Male indistinguishable in the field from Purple Martin although white feathers on lower belly are visible in specimens. Female has brown throat, breast, and sides and unstreaked belly. Juvenile similar to female.

SIMILAR SPECIES. (1) Male Purple Martin is virtually indistinguishable, but female has streaked belly. (2) Black Swift is dark brown, with narrower and longer wings and faster flight.

RANGE. Cuba. Winter range unknown, presumably South America. **Status**: Common summer resident in Cuba, Isla de Pinos, and Cayo Romano (27 Jan–21 Oct). **Habitat**: Open country, towns. **Nesting**: Apr–Aug. Colonial, but scattered pairs may breed alone. Lays three to five white eggs in abandoned woodpecker holes or natural tree cavities; also in old buildings. **Voice**: Very similar to that of Purple Martin. **Food**: Insects caught on the wing.

TREE SWALLOW

GOLONDRINA DE ÁRBOLES

Tachycineta bicolor   Pl. 36

DESCRIPTION. 5.5" (14 cm). *Male*: upperparts dark metallic greenish blue, appearing black from a distance; white below. Tail slightly forked. *Female*: resembles male, usually slightly duller. In both sexes, the wings appear brown or blackish brown in flight. Juvenile grayish brown above; white below. Feeds in huge flocks at times; low above the ground.

SIMILAR SPECIES. Bahama Swallow has violet blue gloss on wings, rump, and tail, which is deeply forked.

RANGE. Breeds in North America, wintering south to Honduras, occasionally farther; Bahamas and Greater Antilles, including the Cayman Islands. **Status**: Abundant winter resident in Cuba and Isla de Pinos (Nov–21 May). Also summer records of transients (3 Jun–4 Sep). **Habitat**: Swamps, rice fields, open country. **Voice**: A twittering *klweet*. **Food**: Insects caught on the wing.

BAHAMA SWALLOW

GOLONDRINA DE BAHAMAS

Tachycineta cyaneoviridis   Pl. 36

DESCRIPTION. 6" (15 cm). Dark dull green above, with violet blue gloss on wings, rump, and tail; white below. Tail deeply forked. Feeds in flocks; flight very rapid.

SIMILAR SPECIES. Tree Swallows wintering in Cuba have mostly brown or blackish brown wings, their tails are only slightly forked, and upperparts are glossy greenish blue.

RANGE. Breeds in northern Bahamas, wintering throughout the Bahamas and eastern Cuba. **Status**: Rare winter resident (30 Jan–8 Mar). Vulnerable. **Habitat**: Open fields with low vegetation. **Voice**: A low, continuously repeated *chi-weet*. **Food**: Insects caught on the wing.

NORTHERN ROUGH-WINGED SWALLOW
GOLONDRINA PARDA

Stelgidopteryx serripennis Pl. 36

DESCRIPTION. 5.5" (14 cm). Warm brown above; white below, with a brown wash on the chin, throat, and chest. Tail slightly forked. Juvenile has cinnamon wing bars. Feeds and migrates singly or in small flocks.

SIMILAR SPECIES. Bank Swallow has well-defined brown breast band and shallow and faster wingbeats.

RANGE. Breeds in southern Canada, United States, and northern Mexico, wintering south to Panama. **Status**: Common transient in Cuba and Isla de Pinos, most often seen during spring migration; rare winter resident (13 Aug–12 Apr). **Habitat**: Savannas, swamps, open fields, usually near water. **Voice**: A rich, *brrrtt*. **Food**: Insects caught on the wing.

BANK SWALLOW
GOLONDRINA DE COLLAR

Riparia riparia Pl. 36

DESCRIPTION. 5.25" (13 cm). Grayish brown above, white below, with a wide brown breast band. Tail slightly forked. Juvenile has buff wing bars. On migration, some individuals occur among large flocks of Barn Swallows. Wingbeats shallow and fast.

SIMILAR SPECIES. Northern Rough-winged Swallow is more warmly toned above, lacks breast band, and flies more gracefully, with deeper and slower strokes.

RANGE. Northern Hemisphere, wintering from Panama to South America. **Status**: Rare transient in Cuba and Isla de Pinos (1 Aug–30 Oct; 4 Feb–26 Jun). **Habitat**: Open forests, savannas, often near coasts. **Voice**: A buzzy, incessantly repeated *bijzzz*. **Food**: Insects caught on the wing.

BARN SWALLOW
GOLONDRINA COLA DE TIJERA

Hirundo rustica Pl. 36

DESCRIPTION. 6.75" (17 cm). *Male*: very dark steely blue above; cinnamon below, with a narrow dark-blue breast band. Forehead and throat chestnut. Tail deeply forked, with white marks forming band as seen from below. *Female*: paler below; upperparts with less gloss and shorter tail. Juvenile similar to female but upperparts mixed with brown. Flight is low and swooping.

SIMILAR SPECIES. Both (1) Cave and (2) Cliff Swallows have square tail and buffy rump.

RANGE. Breeds in Eurasia, northern Africa, North America, and Mexico; in the Americas, winters from Panama to South America, Puerto Rico, and Lesser Antilles. **Status**: Common transient in Cuba, Isla de Pinos, and some larger cays (1 Aug–20 Nov; 3 Feb–26 Jun). **Habitat**: Open country, grassy meadows, often near coasts. **Voice**: A soft *tweek*, *slip*, or *kleep-it*. **Food**: Insects caught on the wing.

CLIFF SWALLOW

GOLONDRINA DE FARALLÓN

Petrochelidon pyrrhonota　　　　　　　　　　　　　　　　　　Pl. 36

DESCRIPTION. 5.5" (14 cm). Very dark glossy blue above with buffy orange rump; whitish belly. Sides of neck dark cinnamon; chin and throat blackish. Forehead pale. Tail square. Juvenile has brownish gray upperparts (except rump); tertials have buffy rufous edging.

SIMILAR SPECIES. (1) Cave Swallow has cinnamon throat and forehead. (2) Barn Swallow has deeply forked tail; lacks buffy rump.

RANGE. Breeds in North America, wintering from Paraguay to central Argentina. **Status**: Rare transient in Cuba and Isla de Pinos (30 Aug–11 Nov; 21 Mar–2 Jun). **Habitat**: Open fields. **Voice**: A short, melodious *chur*, as well as peculiar squeaky notes. **Food**: Insects caught on the wing.

CAVE SWALLOW

GOLONDRINA DE CUEVAS

Petrochelidon fulva　　　　　　　　　　　　　　　　　　　　Pl. 36

DESCRIPTION. 5.5" (14 cm). Very dark glossy blue above with buff rump; white belly. Cinnamon forehead, sides of neck, chin, and throat. Square tail. Juvenile is duller above, with spotted throat.

SIMILAR SPECIES. (1) Cliff Swallow has very dark throat (cinnamon to black) and whitish forehead. (2) Barn Swallow has deeply forked tail and is entirely dark blue above.

RANGE. Texas, New Mexico, Florida, Mexico, Ecuador, and Peru; Greater Antilles. **Status**: Common summer resident in Cuba and Isla de Pinos (29 Jan–21 Oct). **Habitat**: Open fields, savannas, city suburbs. **Nesting**: Mar–Aug. Nest is built mainly of mud, on cliffs, in caves, under bridges or eaves. Lays three or four white eggs, heavily spotted with brown. **Voice**: A melodious *chur* and chattering calls. **Food**: Insects caught on the wing.

WRENS Troglodytidae

Small, rather chunky, and very energetic birds with long legs, short wings, and short tail, usually held cocked. Almost all wrens are brown with barred wings and tail, and thin and slightly decurved bill. Fine, and in many cases truly exceptional, singers. Strongly territorial. Mostly insectivorous. Sexes alike. (**W**:75; **C**:3)

ZAPATA WREN
FERMINIA; FERMINA
Ferminia cerverai   Pl. 35

DESCRIPTION. 6.25" (16 cm). Brown, finely barred with black above and brownish white below. Bill rather long; tail long, usually pointing down when the bird is perched. A poor flier with very short wings. Skulks and is often difficult to observe.

SIMILAR SPECIES. Marsh Wren is smaller, with white stripes on back and distinct white eyebrow.

RANGE. Endemic to Cuba. **Status**: Endangered. Restricted to Zapata peninsula where it is known mainly from a small area north of the village of Santo Tomás and near the mouth of the Hatiguanico river, La Yuca, Peralta, Hato de Jicarita, Canal de los Patos. **Habitat**: Marshes with extensive fields of sawgrass and patches of shrubs seasonally flooded to a depth of about 0.5 m. **Nesting**: Mar–May. Nest is globular with a side entrance; known to lay at least two eggs, but few nests have been observed. **Voice**: Male song is a pleasant, loud, canary-like warbling beginning with one to three sweet whistled introductory notes followed by a complex series of grating rattles and buzzes, often repeated in pairs or triplets. Call notes, usually given by the female, include a guttural *kraok* and often repeated metallic *tik* notes not unlike two stones clicked together. **Food**: Insects, spiders, small fruits, mollusks, lizards.

HOUSE WREN
TROGLODITA AMERICANO

Troglodytes aedon   Pl. 35

DESCRIPTION. 4.75" (12 cm). Brown above; somewhat paler below. Faint pale eyebrow. Wings, sides, and tail finely barred with black.

SIMILAR SPECIES. Marsh Wren has conspicuous eyebrow and stripes on back.

RANGE. Breeds in North America, Central America, and virtually throughout South America. Most North American breeders winter from southern United States to southern Mexico. **Status**: Vagrant. One record: 19 Jan (1964). Coastal scrub in La Habana. **Habitat**: Thickets, shrubbery, often near human habitation. **Voice**: A loud and long, rising and falling bubbling whistle. **Food**: Insects.

MARSH WREN*
TROGLODITA DE CIÉNAGA

Cistothorus palustris   Pl. 51

DESCRIPTION. 5" (13 cm). Rich brown above, with conspicuous white stripes on back and a distinct white eyebrow. Mostly whitish below; cinnamon sides. Juvenile similar to adult but duller.

SIMILAR SPECIES. (1) House Wren has entirely brown back and faint eyebrow. (2) Zapata Wren is larger with dark bars on back rather than white stripes, and a less distinct eyebrow.

RANGE. Breeds in North America; also locally in the state of México. Winters to central Mexico. **Status**: A single recent sighting in Los Canales, Zapata peninsula: 8 Aug (1993). **Habitat**: Marshes, especially with cattails. **Voice**: A gurgling splutter; also a loud *tsuk*. **Food**: Aquatic insects.

KINGLETS Regulidae

Very small and delicate birds. All species are greenish, and have prominent wing bars and colorful crown patches that are displayed while singing or in aggressive encounters. All are arboreal and insectivorous. (**W**:5; **C**:1).

RUBY-CROWNED KINGLET
REYEZUELO

Regulus calendula   Pl. 37

DESCRIPTION. 4.5" (11 cm). Very small and plump, appearing almost neckless. Olivaceous gray above and pale yellowish gray below, with a prominent eye ring and white wing bars. Male has concealed scarlet crown patch. Bill very short and thin.

SIMILAR SPECIES. Some vireos and warblers have basically similar color patterns, but are larger and proportionally slimmer, with longer and heavier bills.

RANGE. Breeds in northern and western North America, wintering south to Guatemala. **Status**: Vagrant. One record: 18 Oct (1964). Near La Habana. **Habitat**: Woods, thickets. **Voice**: A fast, sharp *ji-dit*. **Food**: Insects, spiders.

GNATCATCHERS Sylviidae

These small birds are generally drab, in browns and grays, but there are exceptions. Many are fine singers and most are strictly insectivorous. Species occurring in Cuba are tiny and arboreal. Nests are cup shaped or globular, found in bushes or trees. Sexes different in many species. (**W**:552; **C**:2)

BLUE-GRAY GNATCATCHER
RABUITA
Polioptila caerulea   Pl. 37

DESCRIPTION. 4.5" (11 cm). Much like Cuban Gnatcatcher, but with a slightly larger white eye ring and without black crescent. Males in late winter have a distinct thin black eyebrow, otherwise, males and females lack black facial markings during winter.

SIMILAR SPECIES. Cuban Gnatcatcher has black auricular crescent, and louder and more varied vocalizations.

RANGE. Southeastern Canada and United States to Guatemala, wintering to Honduras, Bahamas, Cuba, and Cayman Islands. **Status**: Common winter resident on Cuba, Isla de Pinos, and some larger cays (10 Aug–1 May). **Habitat**: Tall forests, gardens. **Voice**: Common call note is a buzzy *speeee*, almost identical to Cuban Gnatcatcher. Sings from February in Cuba. Song is an insect-like, thin, high warble. **Food**: Small insects.

CUBAN GNATCATCHER
SINSONTILLO
Polioptila lembeyei   Pl. 37

DESCRIPTION. 4.25" (11 cm). Bluish gray above; white below. Tail long, black, with white outer feathers. A narrow black crescent in auricular area, usually broader in males. Juvenile is paler, with a wash of olive on back; the crescent barely evident. Slim and restless. Tail frequently cocked.

SIMILAR SPECIES. Wintering Blue-gray Gnatcatchers generally lack black facial markings, although late-winter males show black eyebrow. Vocalizations simpler.

RANGE. Endemic to Cuba. **Status**: Common, but local. Restricted to scattered coastal localities: Casilda, Punta Maisí to Cabo Cruz, Nuevitas, Santa Lucía, the cays north of Camagüey and Ciego de Ávila provinces, near Bayamo, and Batiquirí. **Habitat**: Dry, dense thorn scrub; coastal vegetation. **Nesting**: Mar–Jul. Nest is cup shaped; larger but otherwise not unlike that of hummingbirds. Lays three to five brown-spotted, white eggs. **Voice**: Song, a rather loud and long rambling series of warbles, whistles, and chattering notes *psss-psss, tiizzzzz-tzi-tzii*, etc., disorganized but sustained; also an incessantly repeated *pip* or *pyip*. Mimicked call notes of other species are often placed at beginning of song and may include those of American Kestrel, Northern Flicker, Loggerhead Kingbird, and Red-legged Thrush. Common call note is a buzzy *speeee*, almost identical to Blue-gray Gnatcatcher. **Food**: Small insects, spiders.

THRUSHES Turdidae

Small to medium-sized birds of worldwide distribution. Bill pointed; legs generally long. Plumage color is highly variable; in many species, adults and juveniles are spotted below. Food is mainly fruits and insects. Thrushes are commonly ground dwellers, and many are extraordinarily gifted singers. Most species are not strongly sexually dimorphic. (**W**:179; **C**:11)

NORTHERN WHEATEAR

TORDO ÁRTICO

Oenanthe oenanthe   Pl. 37

DESCRIPTION. 5.75" (15 cm). *Male*: in breeding plumage, gray above and cinnamon buff below. Wings, broad terminal tail band, and cheeks black. *Female*: similar to male, with brownish upperparts. Wings, terminal tail band, and stripe across eye brown. Both sexes have large white rump patch extending to the base and sides of tail. Male in fall resembles female. Immature is cinnamon buff below, with buff eyebrow.

SIMILAR SPECIES. None.

RANGE. Breeds in North American Arctic, northern Europe, Asia, and northern Africa. Winters in Africa and Asia. Casual along United States eastern seaboard in migration. **Status**: Vagrant. One record: 16 Oct (1903). Holguín province. **Habitat**: Open country. **Voice**: A harsh *chak*. **Food**: Insects, seeds, fruits.

EASTERN BLUEBIRD

AZULEJO PECHIRROJO

Sialia sialis   Pl. 37

DESCRIPTION. 7" (18 cm). *Male*: blue above; rusty red below, with white lower belly and undertail coverts. *Female*: like male, but duller.

SIMILAR SPECIES. None.

RANGE. Breeds in eastern and central North America south to Nicaragua; winters within this range. **Status**: Very rare winter resident in Cuba (30 Nov–30 Apr). **Habitat**: Open forests, orchards. **Voice**: A musical *chur-wi-wee*. **Food:** Insects, fruits, earthworms.

CUBAN SOLITAIRE

RUISEÑOR

Myadestes elisabeth   Pl. 37

DESCRIPTION. 7.5" (19 cm). Slender. Olivaceous brown above; very pale gray below. Rather long tail, usually hanging straight down below the body. A creamy white eye ring, brown whisker stripe, and white-edged outer tail feathers. Juvenile is more olive above; some traces of brown flecks below; tertials have buff tips. Much more readily heard than seen when perching in the higher branches. A superb singer.

SIMILAR SPECIES. La Sagra's Flycatcher has cinnamon brown tail, lacks whisker stripe, and commonly holds tail in line with back.

RANGE: Endemic to Cuba and formerly Isla de Pinos. **Status**: Common but confined to eastern and western mountain ranges. Easily seen at La Güira, Sierra de la Güira; Sierra Maestra; and Cupeyal. **Habitat**: Semi-deciduous woods and pine forests near limestone cliffs. **Nesting**: May–Jul. Nest is a cup constructed of rootlets, hair, and lichens, built in crevices in limestone cliffs as well as in tree cavities; cavity entrance is usually shielded by bromeliads. Lays three pale-green, brown-spotted eggs. **Voice**: The song is a series of sustained, flutelike, almost harsh *zheeee* notes delivered very deliberately on varied pitches, interspersed with briefer warbled or trilled phrases. Not unlike the sound produced by running a wet finger around the rim of a wine glass. Undoubtedly the country's best singer. **Food**: Mainly hovers in trees for insects, fruits, seeds.

VEERY

TORDO COLORADO

Catharus fuscescens   Pl. 37

DESCRIPTION. 7" (18 cm). Uniformly reddish brown above; throat and upper breast very finely spotted and washed with the same color. Belly white. Eye ring indistinct.

SIMILAR SPECIES. (1) Gray-cheeked, (2) Bicknell's, and (3) Swainson's Thrushes have olive brown backs and conspicuous brown spots on chest. (4) Wood Thrush is larger, with conspicuous white eye ring and bold black spots below.

RANGE. Breeds in southern Canada and northern United States, wintering in South America from Colombia to Brazil. **Status**: Rare transient in Cuba (12 Sep–22 Oct; Apr–4 May). **Habitat**: Open forest, gardens. **Voice**: A low whistle *phew* or *veer* and a dry cackle. **Food**: Insects, fruits, spiders.

GRAY-CHEEKED THRUSH

TORDO DE MEJILLAS GRISES

Catharus minimus   Pl. 37

DESCRIPTION. 7.5" (19 cm). Olive brown above. Throat and breast whitish with brown spots, with a buffy wash in some specimens. Belly white. Cheek gray, streaked with white. Inconspicuous eye ring and lores whitish.

SIMILAR SPECIES. (1) Bicknell's Thrush is smaller and browner above, with chestnut tail. Probably not safely differentiated in the field. (2) Swainson's Thrush has conspicuously buffy eye ring, lores, and chest. (3) Veery has throat and breast very finely spotted and washed with pale reddish brown. (4) Wood Thrush is larger, reddish brown above, with large and bold black spots below.

RANGE. Breeds in northern North America and Siberia, wintering south to northern South America. **Status**: Rare transient in Cuba, Isla de Pinos, and some larger cays (14 Sep–28 Nov; 19 Mar–13 May). **Habitat**: Open forests. **Voice**: A nasal *pheu*. **Food**: Insects, fruits.

BICKNELL'S THRUSH

TORDO DE BICKNELL

Catharus bicknelli

Not Illustrated

DESCRIPTION. 7.25" (19 cm). Very similar to Gray-cheeked Thrush, but smaller, browner above, with chestnut tail. Under most circumstances, the two species cannot be safely differentiated in the field.

SIMILAR SPECIES. (1) Gray-cheeked Thrush is larger and more olive above; lacks chestnut tail, but see above. (2) Swainson's Thrush has conspicuously buffy eye ring, lores, and chest. (3) Veery has throat and breast very finely spotted and washed with pale reddish brown. (4) Wood Thrush is larger, reddish brown above, with large and bold black spots below.

RANGE. Northeastern North America, wintering in Hispaniola. **Status**: Very rare transient. Winters in small numbers on Pico Turquino. Two records: Oct (1965, 1968). Botanical Garden of La Habana, Pico Cuba. **Habitat**: Open forest. **Voice**: A nasal *pheu*. **Food**: Insects, fruits.

SWAINSON'S THRUSH

TORDO DE ESPALDA OLIVADA

Catharus ustulatus   Pl. 37

DESCRIPTION. 7″ (18 cm). Olive brown above. Upper breast, lores, auricular streaks, and conspicuous eye ring buffy. Breast spotted with brown; belly white.

SIMILAR SPECIES. (1) Gray-cheeked and (2) Bicknell's Thrushes have inconspicuous whitish eye ring and lores; breast is usually white, sometimes pale buffy. (3) Veery is reddish brown above; throat and breast are very finely spotted and washed with the same tone. (4) Wood Thrush is larger, reddish brown above with white underparts covered with bold black spots.

RANGE. Breeds in northern and western North America, wintering south to Argentina. **Status**: Rare transient in Cuba, Isla de Pinos, and some larger cays (11 Sep–28 Nov; 19 Mar–10 May). **Habitat**: Open forests, bushy areas, gardens. **Voice**: A short *beep*. **Food**: Insects, fruits.

HERMIT THRUSH

TORDO DE COLA COLORADA

Catharus guttatus

Not Illustrated

DESCRIPTION. 7″ (18 cm). Brownish olive upperparts with contrasting rufous tail. Whitish eye ring. White throat. Upper breast densely smudged with dark brown spots. Lower breast and belly white. Immatures often have buff-tipped greater secondary coverts, forming single wing bar.

SIMILAR SPECIES. (1) Veery is wholly reddish brown above; breast is only faintly spotted. (2) Swainson's and (3) Gray-cheeked Thrushes lack contrasting rufous tail. (4) Bicknell's Thrush has only faintly reddish brown tail and lacks whitish eye ring.

RANGE. Breeds throughout northern and western United States and southern Canada north to Alaska, wintering from western and southern United States to Guatemala and El Salvador. **Status**: Vagrant. One record: 25 Dec

(1995). Cayo Coco. **Habitat**: Forests and thickets. **Voice**: A short *yurp* and a whining *weeeen*. **Food**: Insects, fruits.

WOOD THRUSH
TORDO PECOSO
Hylocichla mustelina   Pl. 37

DESCRIPTION. 8" (20 cm). Bright reddish brown above, the color being more intense on head. White below, extensively marked with very bold black spots. A conspicuous white eye ring.

SIMILAR SPECIES. (1) Veery is smaller with very inconspicuous markings below and weak eye ring. (2) Swainson's, (3) Gray-cheeked, and (4) Bicknell's Thrushes are smaller, lack bright rusty tones, and are less extensively and boldly marked below.

RANGE. Breeds throughout eastern United States and southeastern Canada, wintering from Texas to Panama. **Status**: Rare transient and possibly extremely rare winter resident in Cuba (14 Sep–21 Nov; 8 Feb–18 Apr). **Habitat**: Mature semideciduous forest, gardens. **Voice**: A snappy *wit-wit-wit-wit* and a lower *tut-tut-tut-tut*. **Food**: Insects, fruits.

AMERICAN ROBIN
ZORZAL MIGRATORIO
Turdus migratorius   Pl. 38

DESCRIPTION. 10" (25 cm). *Male*: dark gray above with blackish head and bold broken white eye ring. Underparts reddish brown; lower belly white; head blackish. Bill yellow. *Female*: much duller overall.

SIMILAR SPECIES. Red-legged Thrush has orange red bill, legs, and eye ring and only a slight wash of reddish brown on lower belly.

RANGE. North America and Mexico, wintering south to Guatemala and rarely the northern Bahamas. **Status**: Very rare transient to Cuba, sometimes occurring in the western half of the island after the passage of a fall or winter cold front (23 Sep–21 Nov; 1 Mar–18 Apr). **Habitat**: Open country. **Voice**: A staccato *tut-tut-tut*. In flight, a thin *see-lip*. **Food**: Insects, fruits.

RED-LEGGED THRUSH
ZORZAL REAL; ZORZAL DE PATAS COLORADAS
Turdus plumbeus   Pl. 38

DESCRIPTION. 10.5" (27 cm). Slate gray with a slight blue cast. Eye ring and legs orange red; bill dark orange becoming dark in Feb–June; throat black. Lower belly gray in eastern populations, reddish brown in birds from central and western Cuba and Isla de Pinos. Outer tail feathers tipped with white. Juvenile is duller overall with gray throat, spotted with black. Feeds on the ground among leaf litter.

SIMILAR SPECIES. American Robin has entirely reddish brown chest and belly, yellow legs, and bill and white eye ring.

RANGE. Bahamas, Cuba, Cayman Brac, Hispaniola, Puerto Rico, Dominica, and Swan Islands. **Status**: Common permanent resident on Cuba, Isla de Pinos, and many of the larger cays off northern and southern coasts. **Habitat**: Semideciduous woods, pine forests, cactus scrub, gardens. **Nesting**: Mar–Nov. Builds bulky nest of grasses and dry leaves, lined with hairs, feathers, and plant fibers. Lays three to five greenish-white eggs, spotted with brown. **Voice**: A series of creaks and whistles, commonly uttered in pairs with a distinct pause between each note *chirruit*-(pause)-*chirruit* or *pert*-(pause)-*squeeer* or *squit*-(pause)-*seeer*. When alarmed and taking wing, a *wet-wet*. Also a mewing note reminiscent of Gray Catbird. Often incorporates call notes of other species into its song, including those of Black-necked Stilt, Red-tailed Hawk, Northern Flicker, Yellow-throated Vireo, Oriente Warbler, Cuban Bullfinch, and Red-legged Honeycreeper. **Food**: Small insects, lizards, and fruits, commonly those of royal palm.

MOCKINGBIRDS AND THRASHERS Mimidae

Medium-sized birds with long tails. Most species are largely brown or gray. Their voices are loud and varied and some are accomplished singers and mimics. Their diet includes fruits, insects, and small lizards. Sexes alike. (**W**:74; **C**:4)

GRAY CATBIRD
ZORZAL GATO
Dumetella carolinensis   Pl. 38

DESCRIPTION. 8.5" (22 cm). Dark slaty gray, with black cap and tail. Undertail coverts reddish chestnut.

SIMILAR SPECIES. None.

RANGE. Southern Canada to Gulf states, wintering south to Panama and Greater Antilles, and Cayman Islands. **Status**: Common winter resident and transient in Cuba, Isla de Pinos, and several larger cays (19 Sep–25

May). **Habitat**: Dense shrubbery, from sea level to mountains. **Voice**: A quiet, somewhat catlike *mew*; also a repeated *turrr* or *tirrr*, a *quert*, and a cackling *kutakut*. **Food**: Fruits, insects.

NORTHERN MOCKINGBIRD

SINSONTE

Mimus polyglottos Pl. 38

DESCRIPTION. 10″ (25 cm). Dull gray above and whitish below, with a large white wing patch and white tail margins. Eye yellow. Juvenile is streaked below, with brownish eye. Often extends wings when foraging on ground.

SIMILAR SPECIES. Bahama Mockingbird has whisker mark, streaks on sides, faintly on upper breast, and lacks white wing patch. Restricted to northern cays.

RANGE. North America, Mexico, Bahamas, Greater Antilles, and Little Cayman Island. **Status**: Common permanent resident in Cuba, Isla de Pinos, and many offshore cays. **Habitat**: Villages, gardens, agricultural areas with shrubs, coastal vegetation. **Nesting**: Feb–Aug. Nest is a bulky cup of coarse dead twigs, weed stems, decayed leaves, rags, string, and cotton, lined with fine grasses, rootlets, and hair, placed in dense bushes or low in trees. Lays three pale bluish-green eggs, spotted and blotched with brown. **Voice**: Varied, loud, and melodious phrases; many are imitations of other birds' calls or mechanical sounds. A common call is a harsh *tchek*. Male's delightful song in breeding season may be sustained for tens of minutes and is often delivered at night. **Food**: Insects, small fruits. Forages mainly on the ground.

BAHAMA MOCKINGBIRD

SINSONTE PRIETO; SINSONTE CARBONERO

Mimus gundlachii Pl. 38

DESCRIPTION. 11″ (28 cm). Brownish gray above; paler below. Back, upper breast, and sides sparingly streaked with dark brown. Throat bordered by two black whisker stripes. Wing lacks white patch; tail tipped, but not bordered, with white. Juvenile is densely spotted below.

SIMILAR SPECIES. Northern Mockingbird is slightly smaller with conspicuous white wing patch and unstreaked underparts.

RANGE. Bahama Islands, Cuba, and Jamaica. **Status**: Rare and vulnerable permanent resident on the larger cays north of Camagüey and Ciego de Ávila provinces; common on Cayos Cruz and Guillermo. **Habitat**: Low shrubbery near coast with grass, palmettos, and scattered trees. **Nesting**: Apr–Jul. Builds nest among tall bushes. Usually lays three creamy-white, spotted eggs. **Voice**: Song abrupt and loud, but far less melodious and varied than that of Northern Mockingbird, and not incorporating mimicry. Often begins with a characteristic whistled *sereee* note. Call is a *tyerrp*, longer and slightly higher pitched than Northern Mockingbird. **Food**: Insects, small fruits.

BROWN THRASHER

SINSONTE COLORADO

Toxostoma rufum   Pl. 38

DESCRIPTION. 11.5″ (29 cm). Rufous above with long rufous tail; creamy white below, heavily streaked with dark brown. Eye yellow. Wing coverts have black bars and buff tips, forming a pair of wing bars. Immature has brown or gray eye.

SIMILAR SPECIES. None.

RANGE. Southern Canada and central and eastern United States, wintering within the United States. **Status**: Vagrant. Three records: 29 Sep; 4 Oct (1963); 15 Oct. La Habana. **Habitat**: Gardens, agricultural areas with shrubs, coastal vegetation, open, low forest with dense shrubbery. **Voice**: A smacking *tsuk*. **Food**: Insects, fruits.

STARLINGS Sturnidae

This large Old World family comprises mainly medium-sized, robust birds; many have short, squared tails and stocky, strong legs and bills. Most members of the family have glossy plumage with a metallic sheen; generally the sexes are quite similar. Often gregarious. (**W**:148; **C**:1)

EUROPEAN STARLING

ESTORNINO

Sturnus vulgaris Pl. 37

DESCRIPTION. 8.5" (22 cm). Winter plumage blackish and speckled, with green and violet sheen. Bill long, straight, and dark. Breeding adults have yellow bill and iridescent black plumage. Immature is dusky brown with brown bill. Feeds on the ground.

SIMILAR SPECIES. None.

RANGE. Eurasia. Introduced in North America during the 19th century and now widespread there. **Status**: Vagrant. Three records: two on 11 Nov (1964); one undated. Near Gibara in Holguín province and in La Habana. **Habitat**: Cities, towns, agricultural fields. **Voice**: Varied and mostly pleasant squeaks, whistles, warbles. A skillful mimic. **Food**: Insects, seeds.

WAXWINGS Bombycillidae

Medium-sized birds with short, rather wide bills and an elegant, long, back-pointing crest. Wings and sometimes tail may be decorated with small bright red waxy tips. Highly frugivorous and hence highly social. Sexes alike. (**W**:8; **C**:1)

CEDAR WAXWING

PICOTERO DEL CEDRO

Bombycilla cedrorum Pl. 37

DESCRIPTION. 7" (18 cm). Silky and brownish overall, with pale yellow belly. Tail gray, shading to black and ending in bold yellow band. Tips of secondaries red in most adults. A conspicuous black mask and a rather short bill. The long crest is commonly compressed against the head. Juvenile is streaked below, with pale throat and yellow-tipped tail.

SIMILAR SPECIES. None.

RANGE. Breeds in southern Canada and northern United States, wintering south to Panama, Bahamas, Greater Antilles, and Cayman Islands. **Status**: Common transient and winter resident in Cuba, Isla de Pinos, San Felipe, and Cayo Coco (14 Oct–22 May). **Habitat**: Semideciduous woods, pine forests, trees or bushes with berries. **Voice**: A thin, high-pitched, slightly trilled *ssseeeee*. **Food**: Small fruits, flying insects.

WOOD-WARBLERS Parulidae

Small, brightly colored insectivorous birds with short, thin bills. Some species dwell among the highest branches; others, in bushes; and still others, on or very near the ground. This is a very homogeneous group, with many species having similar color patterns; thus occasionally they are difficult to differentiate in the field. North American species are sexually dimorphic. Generally solitary, some species are social to the extent that they join flocks of other birds on their Cuban wintering grounds. Many species that breed in North America migrate as far south as South America, passing through either Central America or the West Indies. Many warblers found in Cuba in fall and winter are young birds of the previous season, and though their pattern suggests that of the adults, their plumage is often duller. Generally, the warblers wintering in Cuba do not sing here, uttering only call notes, some of which are quite distinctive. However a few sing until October, but mostly they can be heard singing from mid-March. The earliest spring migrant is the Louisiana Waterthrush, which can be heard singing from January. Popularly, the name *chinchillita* is used for all warblers. (**W**:115; **C**:41)

BACHMAN'S WARBLER

BIJIRITA DE BACHMAN; BIJIRITA DE PECHO NEGRO

Vermivora bachmanii   Pl. 39

DESCRIPTION. 4.5" (11 cm). *Male*: olive above, grayish on nape, with black crown and yellow forehead. Yellow below, with a large black patch on throat and breast. Lower belly and undertail coverts white. Tail with white spots. Eye ring yellow. Lacks wing bars. Bill long, slender, and distinctly decurved. *Female*: similar, without black markings but may have a faint gray patch on breast. Immature duller.

SIMILAR SPECIES. (1) Male Hooded Warbler has black hood with large yellow mask. Often flicks and fans open tail. (2) Oriente Warbler has all-gray upperparts, white belly, and unmarked tail. (3) Female Wilson's Warbler has olive head above and is entirely yellow below; tail is unmarked.

RANGE. Breeds, or bred, in southeastern United States, wintering in Cuba. **Status**: Winters on Cuba and Isla de Pinos (7 Sep–16 Mar). Critically endangered. Extremely rare if not extinct; last confirmed records in 1962 and 1964 at Zapata peninsula. **Habitat**: Swamp forest. **Voice**: Short buzzy notes rapidly repeated in series on one pitch. **Food**: Insects.

BLUE-WINGED WARBLER

BIJIRITA DE ALAS AZULES

Vermivora pinus   Pl. 40

DESCRIPTION. 4.75" (12 cm). *Male*: bright yellow head and underparts, with olive nape and back, and gray wings with white bars. A black eye line. *Female*: quite similar, with olive forehead and crown. Both sexes have white spots in tail. Immature male has pale yellow forehead and crown, gradually blending into the color of the back. Eye line dull black. Immature female has olive forehead with a diffuse yellow patch; eye line dusky.

SIMILAR SPECIES. (1) Prothonotary Warbler has unmarked wings and head. (2) Female Golden-winged Warbler is white below, with gray throat. Wing marked with yellow. Hybrids are known to visit Cuba. (3) Yellow Warbler has olivaceous wings with yellow wing bars and yellow tail spots, and lacks dark eye line.

RANGE. Breeds in eastern North America, wintering from Mexico to Panama. **Status**: Rare winter resident and transient in Cuba, Isla de Pinos, and some larger cays (21 Aug–12 Apr). **Habitat**: Thickets and bushes. Also forests, feeding in middle and upper branches. **Voice**: A sharp, loud *jeet*. Primary song a buzzy *beeee-buzzzz*. **Food**: Insects.

GOLDEN-WINGED WARBLER

BIJIRITA ALIDORADA

Vermivora chrysoptera Pl. 40

DESCRIPTION. 5" (13 cm). *Male*: gray above; grayish white below. Crown golden yellow; throat and wide ear patch black. A conspicuous golden yellow patch on wing. *Female*: duller. Both sexes have white tail spots. Immature similar to adult.

SIMILAR SPECIES. Blue-winged Warbler is yellow below, with a black eye line and white wing bars. Interbreeds with Golden-winged Warbler, resulting in hybrids with mixed characteristics.

RANGE. Breeds in northeastern North America, wintering from Yucatán peninsula to northern South America and Greater Antilles. **Status**: Very rare transient in Cuba (1–24 Sep; Feb–7 May). Reported only once during winter at Habana. **Habitat**: Open forests and semideciduous woods. **Voice**: A sharp, loud *jeet*. **Food**: Insects.

TENNESSEE WARBLER

BIJIRITA PEREGRINA

Vermivora peregrina Pl. 39

DESCRIPTION. 4.75" (12 cm). Small and plain. *Male*: winter plumage bright olive green above; pale yellow below with white undertail coverts. Thin but distinct white eyebrow, and a dark eye line. In breeding plumage, head gray, underparts white. *Female*: winter plumage similar to winter male; underparts slightly yellower. Breeding females are similar to breeding male, with a yellow wash on chest. Both sexes have very thin bill, very indistinct wing bars, and unmarked tail; some individuals may have grayish spots on tail. Juvenile similar to winter adult, with somewhat more distinct yellow wing bars. An active warbler, foraging in treetops and calling frequently.

SIMILAR SPECIES. (1) Orange-crowned Warbler has yellow undertail coverts and is faintly streaked below. (2) Philadelphia Vireo has heavier bill, bluish gray (not blackish) feet, and much more deliberate movements. Much less active while foraging.

RANGE. Breeds in Canada and northeastern United States, wintering to northern South America. **Status**: Uncommon transient in Cuba and Isla de Pinos (14 Sep–14 Nov; 8 Feb–7 May). **Habitat**: Forests, gardens. **Voice**: Frequently uttered thin notes: *tsit* or *tsee*. **Food**: Insects.

ORANGE-CROWNED WARBLER

BIJIRITA DE CORONILLA ANARANJADA

Vermivora celata   Pl. 40

DESCRIPTION. 5" (13 cm). Olive above; paler below, lightly and dimly streaked. Eyebrow and faint eye ring yellowish. Undertail coverts yellow; tail unmarked. A concealed orange patch on crown, very seldom visible. Female similar to male, but with a smaller crown patch (noticeable only in the hand). Immature has grayish head and whitish eye ring and eyebrow.

SIMILAR SPECIES. (1) Tennessee Warbler is unstreaked below, with white undertail coverts. (2) Nashville Warbler is bright unstreaked yellow below, with a white patch on lower belly and a more conspicuous eye ring.

RANGE. Breeds in northern and western North America, wintering from southern United States to Guatemala. **Status**: Vagrant. Three records: 11 Nov (1989); 21 Oct (1994); 29 Nov (1996). Guanahacabibes peninsula, Cayo Santa María, and Cayo Coco. **Habitat**: Low, coastal vegetation, often foraging near the ground. **Voice**: A *chip*. **Food**: Insects.

NASHVILLE WARBLER

BIJIRITA DE NASHVILLE

Vermivora ruficapilla   Pl. 40

DESCRIPTION. 4.75" (12 cm). Olive green above with gray head and a concealed chestnut crown patch. Yellow below, with a small white area on lower belly. A conspicuous white eye ring. Wing and tail olivaceous, unmarked. Female is duller. Immature duller than female, sometimes with faint brown wash.

SIMILAR SPECIES. (1) Female Mourning and (2) male Connecticut Warblers (not yet known from Cuba, but a possible transient) have entirely gray hood including chin and throat, and have solid yellow belly and pink feet. (3) Female Northern Parula has wing bars and white belly. (4) Orange-crowned Warbler has faint streaking on breast and much less distinct eye ring.

RANGE. Breeds in northern North America, wintering south to Honduras. **Status**: Vagrant. Four records: 9 Oct (1963); 21 Dec (1995); 8 Nov (1996); 20 Jan (1999). La Habana province, Zapata, and Sora. **Voice**: *Jeet*. **Food**: Insects.

NORTHERN PARULA

BIJIRITA CHICA; BIJIRITA PARULA

Parula americana   Pl. 42

DESCRIPTION. 4.5" (11 cm). Very small. *Male*: bluish gray above, tipped green in fall, with a greenish yellow patch on back, two distinct white wing bars, and a prominent broken white eye ring. Chin, throat, and chest are yellow, with narrow black and reddish bands across chest. Belly and tail spots white. *Female*: similar but duller. Never has black chest band but adults may have a faint yellow or orange yellow band on chest. Immature washed green above. A very active feeder, gleaning insects from leaves, often hanging upside down.

SIMILAR SPECIES. (1) Nashville Warbler has plain wings and yellow belly. (2) Yellow-throated Warbler has head and sides boldly marked with black and white.

RANGE. Breeds in eastern North America, wintering in Central America and West Indies. **Status**: Common winter resident and transient in Cuba, Isla de Pinos, and several cays (31 Jul–14 May). **Habitat**: Forests, where it usually forages in the higher branches. **Voice**: A soft, distinct *chip*, frequently repeated. Song a buzzy *zzzzzzzzeeeurp*, often heard in Cuba in late winter. **Food**: Insects, fruit.

YELLOW WARBLER

CANARIO DE MANGLAR; BIJIRITA AMARILLA

Dendroica petechia   Pl. 39

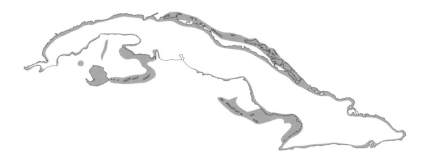

DESCRIPTION. 5" (13 cm). Strikingly yellow under most circumstances. *Male* (resident): yellow orange head with an inconspicuous tawny-reddish cap. Underparts yellow orange; back, wings, and tail olivaceous. Wing bars and inner webs of tail feathers yellow. Streaked with rufous below. *Female* (resident): similar, but with olivaceous head above and plain yellow underparts with faint streaks. Immature is entirely gray, with yellow in primary edges and tail. Wintering birds from North America appear virtually all yellow, and lack the contrasting olivaceous and orange tones of the residents.

SIMILAR SPECIES. (1) Female Hooded Warbler has white tail spots, unmarked underparts, and wings. (2) Female Wilson's Warbler has unmarked tail and wings, and plain yellow underparts. (3) Prothonotary and (4) Blue-winged Warblers have gray wings and tail; the latter has a black eye line and white wing bars.

RANGE. Breeds in most of North and Central America, northern South America, and West Indies. Northern populations winter south to Brazil.

Status: Common permanent resident along the coast of Cuba, Isla de Pinos, and many cays. Also common fall transient in Cuba, Isla de Pinos, and Cayo Coco (15 Aug–10 Oct; 21 Feb–8 May). **Habitat**: Residents live in coastal vegetation, especially mangroves; transient individuals prefer forests and gardens. **Nesting**: Mar–Jul. Builds compact nest in lower branches, lays three or four white, reddish-brown-spotted eggs, with a slight greenish cast. **Voice**: A strong, variable, *tseet-tseet-tee-tee-wee*. **Food**: Insects, spiders.

CHESTNUT-SIDED WARBLER
BIJIRITA DE COSTADOS CASTAÑOS
Dendroica pensylvanica   Pl. 41

DESCRIPTION. 5″ (13 cm). *Male*: winter plumage bright green above, back usually but not always unmarked. Distinct chestnut streak along sides. A narrow prominent white eye ring. Underparts, wing bars, and tail spots white. Breeding male has golden crown, and black eye line and whisker mark surrounding white cheeks. *Female*: winter plumage similar to winter male, but chestnut stripe pale or absent. Breeding plumage is similar to breeding male but duller. Immatures similar to winter female with yellowish wing bars; chestnut sides are almost entirely absent. This species often cocks its tail in a very characteristic manner.

SIMILAR SPECIES. (1) Bay-breasted and (2) Blackpoll Warblers are more or less streaked above and below, appear more olive and yellow than green and white, and lack Chestnut-sided Warbler's prominent eye ring.

RANGE. Breeds in Canada and eastern United States, wintering to eastern Panama, casually to northern South America. **Status**: Uncommon transient in Cuba, Isla de Pinos, and some larger cays (3 Sep–31 Oct; 17 Feb–11 May). **Habitat**: Gardens, forests, from midlevel to canopy. **Voice**: A *chip*. **Food**: Insects, fruits.

MAGNOLIA WARBLER
BIJIRITA MAGNOLIA
Dendroica magnolia   Pl. 39

DESCRIPTION. 5″ (13 cm). *Male*: in winter plumage, gray crown and nape; grayish olive back streaked with black; and upper breast crossed by pale gray band. The rump is conspicuously yellow and there are two white wing bars. Underparts yellow, with sides streaked with black. Eyebrow whitish and thin. Breeding males have broad black mask, black back, and a bold white patch on wing. Black streaks form band across upper breast. *Female*: winter plumage similar to winter male, but duller, the streaks on sides being narrower and the wing bars barely noticeable. Breeding female is much like winter male. Immature is similar to winter female but may be almost completely unstreaked below. Note that both breeding and winter plumages are highly variable, often making distinction of sexes difficult. In all plumages, the tail pattern formed by large white spots in center of all but central feathers is diagnostic, and easily seen as the birds frequently fan the tail.

SIMILAR SPECIES. (1) Cape May Warbler is densely streaked below; tail lacks Magnolia's strong pattern. (2) Yellow-rumped Warbler is mostly whitish below. (3) Prairie Warbler always shows at least faint streaking on face, and wags tail.

RANGE. Breeds in northern North America, wintering throughout Central America and West Indies. **Status**: Common winter resident and transient in Cuba, Isla de Pinos, and some larger cays (10 Sep–20 May). **Habitat**: Forests, gardens. **Voice**: A distinct nasal *enk*; also a dry, two-syllable *chis-sik*. **Food**: Insects.

CAPE MAY WARBLER

BIJIRITA ATIGRADA

Dendroica tigrina Pl. 39

DESCRIPTION. 5″ (13 cm). Small with short tail and very thin bill. *Male*: in winter plumage, grayish green back with dark streaks; yellow underparts extensively streaked with black. Gray ear patch, often tinged with chestnut. Conspicuous white wing patch. Breeding plumage is much brighter, with a conspicuous chestnut ear patch. *Female*: winter plumage similar to winter male but duller, with a blurred yellow spot on sides of neck and whitish wing bars. Breeding plumage similar to breeding male, without chestnut ear patch and with narrower wing bars. Immature male is olivaceous green above, yellowish below with dense streaking; immature female olive gray almost throughout, with densely and softly streaked underparts. In all plumages, has yellowish rump.

SIMILAR SPECIES. (1) Yellow-rumped Warbler is larger, with a brighter and more sharply defined rump patch. It also has small yellow patches on sides and is less streaked below. (2) Magnolia Warbler is less streaked below, with a conspicuous centrally interrupted white band on black-tipped tail. (3) Prairie Warbler has streaks restricted to sides, and more or less bold eye line and mustache, and wags tail frequently. (4) Palm Warbler has yellow undertail coverts and also wags tail.

RANGE. Breeds in northern North America, wintering almost entirely in West Indies. **Status**: Common winter resident and transient in Cuba, Isla de Pinos, and several cays (26 Sep–17 May). **Habitat**: Forests, gardens. **Voice**: A very thin, short *tsit*, easily confused with Stripe-headed Tanager. **Food**: Insects, nectar, fruit.

BLACK-THROATED BLUE WARBLER

BIJIRITA AZUL DE GARGANTA NEGRA

Dendroica caerulescens Pl. 42

DESCRIPTION. 5.25″ (13 cm). *Male*: strikingly patterned. Dark grayish blue above, with a small white spot on wing at base of primaries; white below.

Cheeks, throat, and sides black. *Female*: dark olivaceous brown above; pale brown below with narrow whitish eyebrow and brownish cheek patch; wing spot smaller than in male. Immature male has green-washed upperparts and white-tipped throat feathers. Immature female has wing patch indistinct or occasionally lacking.

SIMILAR SPECIES. Male Cerulean Warbler has white throat and both sexes have wing bars.

RANGE. Breeds in northeastern North America, wintering almost entirely in Bahamas, Greater Antilles, and Cayman Islands. **Status**: Common winter resident and transient in Cuba, Isla de Pinos, and some larger cays (1 Sep–14 May). **Habitat**: Forests and gardens. Feeds from ground level to high in the canopy, but most often seen on lower branches. **Voice**: A dull *chip* or *tip*, like a water drop hitting a leaf. **Food**: Insects, fruits.

YELLOW-RUMPED WARBLER

BIJIRITA CORONADA

Dendroica coronata   Pl. 42

DESCRIPTION. 5.5" (14 cm). *Male*: winter plumage grayish brown above, streaked black, with small yellow patches on crown and side and bright yellow rump. Two very inconspicuous wing bars. Whitish eyebrow and prominent broken eye ring. White below with black streaks on breast. Breeding male blue gray above with black breast band, sides, and auricular patches; throat and belly white. Crown patch yellow. *Female*: winter plumage similar to winter male but duller, with smaller patches of yellow on sides. Breeding plumage resembles winter male.

SIMILAR SPECIES. (1) Immature Cape May Warbler has diffuse yellowish green rump, much less sharply defined, and shorter tail. (2) Palm Warbler has prominent yellow undertail coverts and wags tail frequently. (3) Magnolia Warbler is mostly yellow below and tail has bold black band at tip.

RANGE. The only subspecies known from Cuba is *D. c. coronata*, which breeds in northern and eastern North America, wintering to Panama and West Indies. **Status**: A winter resident and transient in Cuba, Isla de Pinos, and some larger cays (18 Sep–1 May). Common in some years and completely absent in others. **Habitat**: Open country, swamps, gardens. **Voice**: A loud *check* or *chup*. **Food**: Insects.

BLACK-THROATED GRAY WARBLER

BIJIRITA GRIS DE GARGANTA NEGRA

Dendroica nigrescens

Not Illustrated

DESCRIPTION. 5" (13 cm). *Male*: adult plumage essentially the same year-round. Gray above and white below, with prominent black throat and breast giving way to black streaks on sides. Black head boldly marked with white stripes. Yellow loral spot. *Female*: similar to male but duller above, with throat scaled blackish and less black on breast. Immature male like female. Immature female has little black on throat and breast, and is brownish gray above.

SIMILAR SPECIES. Black-and-white Warbler has central white coronal stripe, streaked undertail coverts, and lacks yellow loral spot. Note also creeping behavior.

RANGE. Southwestern Canada and western United States, wintering from California and southwestern United States to south central Mexico. **Status**: Vagrant. One record: 17 Oct (1997). Cayo Coco. **Habitat**: Coastal vegetation, gardens. **Voice**: Call a simple *tup*. **Food**: Insects.

BLACK-THROATED GREEN WARBLER

BIJIRITA DE GARGANTA NEGRA

Dendroica virens   Pl. 42

DESCRIPTION. 5″ (13 cm). *Male*: bright olive green above with mainly yellow cheeks. Chin, throat, and upper breast black; sides streaked with black. Two white wing bars. *Female*: duller, with yellow chin and throat. Immature male resembles adult female; immature female is duller still, without dark markings below apart from dim streaks on sides.

SIMILAR SPECIES. Female and immature Blackburnian Warbler lacks black markings on throat and breast, and have darker and streaked back, and dark area centrally in cheeks.

RANGE. Breeds in Canada and eastern United States, wintering to northern South America and the West Indies, particularly the Bahamas and Greater Antilles. **Status**: Uncommon winter resident and transient in Cuba, Isla de Pinos, and some larger cays (12 Sep–6 May). **Habitat**: Forests, gardens. **Voice**: A short, crisp *tsip*. **Food**: Insects.

BLACKBURNIAN WARBLER

BIJIRITA BLACKBURNIANA

Dendroica fusca   Pl. 39

DESCRIPTION. 5″ (13 cm). *Male*: winter plumage blackish above with pale streaks. Two bold white wing bars. Cheeks have central triangular black patch surrounded by yellowish orange. Chin, throat, and breast yellowish orange. Belly creamy white. Sides streaked with black. Breeding male is similar but more intensely orange. *Female and immature*: similar to winter male but decidedly duller, with yellow replacing orange.

SIMILAR SPECIES. (1) Yellow-throated Warbler has white eyebrow and unmarked gray back. (2) Female Black-throated Green Warbler has broad black band across breast, and immature female Black-throated Green Warbler has unstreaked back and lacks pronounced central dark area in cheek patch.

RANGE. Breeds in northeastern North America, wintering from Costa Rica to Bolivia. **Status**: Rare transient in Cuba, Isla de Pinos, and Cayo Coco (9 Aug–14 Nov; 8 Feb–24 May). **Habitat**: Forests, gardens. **Voice**: A rich *jeet*. **Food**: Insects.

YELLOW-THROATED WARBLER
BIJIRITA DE GARGANTA AMARILLA
Dendroica dominica   Pl. 42

DESCRIPTION. 5.25" (13 cm). Strikingly patterned and with deliberate creeping habits. Gray above with black crown and triangular black ear patch bordered with white, and two white wing bars. Throat and upper breast yellow; belly white. Sides streaked with black; bill rather long. Sexes very similar, with female slightly duller with grayer crown. Immature may be almost imperceptibly washed with brown on back. "Sutton's" Warbler—a hybrid between the Yellow-throated Warbler and Northern Parula—has been reported only once.

SIMILAR SPECIES. (1) Blackburnian Warbler has triangular black or dark ear patch surrounded by orange or yellow. (2) Olive-capped Warbler has bright yellow throat and upper breast patch bordered with black streaks, and lacks streaks on sides.

RANGE. Breeds in eastern United States and Bahamas, wintering to Panama and Greater Antilles. **Status**: Common winter resident and transient in Cuba, Isla de Pinos, and some larger cays (11 Jul–29 Apr). **Habitat**: Forests, gardens, coconut groves. **Voice**: A loud, dry *clip*. **Food**: Insects, spiders.

OLIVE-CAPPED WARBLER
BIJIRITA DEL PINAR
Dendroica pityophila   Pl. 41

DESCRIPTION. 5" (13 cm). Gray above, with olive green crown and two narrow white wing bars. A yellow patch on throat and upper breast, bordered on sides with black streaks. Yellow patch brighter and black border bolder in male. Belly whitish. Juvenile entirely brown with two indistinct whitish wing bars; some individuals show yellowish cast on chin.

SIMILAR SPECIES. Yellow-throated Warbler lacks olive crown patch and black lower border to yellow throat patch; has black side streaks.

RANGE. Grand Bahama, Abaco, and Cuba. **Status**: Common, but very local, permanent resident on Cuba in Pinar del Río (La Güira and Viñales valley) and the eastern provinces of Holguín and Guantánamo. **Habitat**: Found only in pine forests. **Nesting**: Mar–Jun. Builds a cup-shaped nest fairly high in pines and lays two whitish, spotted eggs. **Voice**: Song, a series of seven to nine whistled watery notes, descending slightly and changing in quality. Call, a frequently heard series *tsip-tsip-tsip*. **Food**: Insects.

PINE WARBLER

BIJIRITA DE PINOS

Dendroica pinus Pl. 41

DESCRIPTION. 5.5" (14 cm). Large with rather long bill. *Male*: unstreaked olive above with conspicuous white wing bars. Throat and breast bright yellow; belly white; sides faintly streaked with olive. *Female*: similar but duller with indistinct streaks on sides. In both sexes, tail seems long due to short undertail coverts. Immature mostly brownish, washed with green on back, with yellowish or buffy underparts. All have black feet and legs. This species moves deliberately, often creeping along branches.

SIMILAR SPECIES. (1) Female Bay-breasted Warbler is streaked with black above; feet and legs are gray. (2) Blackpoll Warblers are streaked with black above, have shorter-looking tails due to longer undertail coverts, and have yellowish feet. (3) Yellow-throated Vireo has yellow spectacles and heavier bill.

RANGE. Breeds in eastern North America, wintering to southern United States. Permanent resident in Bahamas and Hispaniola. **Status**: Vagrant. Five records: 22 Oct (1964); 8 Nov (1965); Mar (1987); 17 Jan (1988); 11 Feb (1998). La Habana, Zapata peninsula. **Habitat**: Pine forests, but in Cuba has been found only in *Casuarina* spp. trees with Palm Warblers. **Voice**: A loud, simple *chip*. **Food**: Insects.

PRAIRIE WARBLER

MARIPOSA GALANA

Dendroica discolor Pl. 41

DESCRIPTION. 5" (13 cm). *Male*: olivaceous above; yellow below, with yellowish wing bars. Bold black streaks on face and sides and distinct chestnut streaks on back. *Female*: somewhat duller. Immature similar to female but unstreaked grayish olive above, with gray wash on head and much reduced facial contrasts. Wags tail.

SIMILAR SPECIES. (1) Palm Warbler is grayish brown above, with whitish eyebrow and faintly streaked whitish underparts with contrasting yellow undertail coverts. (2) Pine Warbler has white belly, and lacks black facial streaks. (3) Winter Magnolia Warbler has grayish olive back streaked with black, yellow rump, and black-tipped tail with white central band.

RANGE. Breeds in eastern North America, wintering mainly in West Indies. **Status**: Common winter resident and transient in Cuba, Isla de Pinos, and the larger cays (20 Jul–14 May). **Habitat**: Open forests, thickets, mangroves, scrub vegetation, gardens. **Voice**: A dry, husky *chip*. An ascending, thin buzzy song *zee, zee, zee, zee, zee, zee, zee, zee, zee*. **Food**: Insects, spiders, fruit.

PALM WARBLER
BIJIRITA COMÚN
Dendroica palmarum   Pl. 41

DESCRIPTION. 5" (13 cm). Winter plumage grayish brown above with darker streaks, indistinct wing bars, and whitish eyebrow. Whitish below, with throat and breast faintly streaked with brown, and yellow undertail coverts. Breeding plumage has reddish brown crown, and yellow chin and throat. Wags tail. Commonly occurs in small flocks. Generally feeds on or near the ground.

SIMILAR SPECIES. (1) Immature Cape May Warbler has white undertail coverts and is more heavily streaked below. (2) Yellow-rumped Warbler has white undertail coverts and bright yellow rump. (3) Prairie Warbler is entirely yellow below, with conspicuous black head stripes and streaked sides.

RANGE. Breeds in northern North America, wintering to Nicaragua, Bahamas, and Greater Antilles. **Status**: Abundant winter resident and transient in Cuba, Isla de Pinos, and many cays (24 Aug–17 May). Most arrive in late Sep. **Habitat**: Open forests, savanna, pastures, gardens, coastal vegetation. **Voice**: A distinctive *tsick*. **Food**: Insects.

BAY-BREASTED WARBLER
BIJIRITA CASTAÑA
Dendroica castanea   Pl. 43

DESCRIPTION. 5.5" (14 cm). *Male*: winter plumage yellowish green above streaked with black. Extensive chestnut on sides. White wing bars. Buff below. Breeding plumage olive gray above, with dark streaks; dark chestnut crown, throat, chest, and sides; and a black mask. A cream-colored patch on side of neck. *Female*: winter plumage similar to winter male but chestnut on sides absent or much reduced. Breeding plumage lacks black on head, showing some chestnut on crown and sides. All plumages usually have buff undertail coverts and gray legs and feet. Immature similar to nonbreeding female, with less distinct streaks on back.

SIMILAR SPECIES. (1) Winter Chestnut-sided Warbler is bright unstreaked green above with conspicuous white eye ring. (2) Blackpoll Warbler is slightly darker, more olive green above, with streaked yellowish underparts, extensive pure-white undertail coverts, shorter-looking tail due to longer undertail coverts, and yellowish feet. (3) Pine Warbler has unstreaked greenish-brown back, and is yellow only from chin to upper belly. Feet and legs are black. **RANGE.** Breeds in northern North America, wintering from Panama to northwestern Venezuela. **Status**: Rare transient in Cuba and Isla de Pinos (26 Sep–8 Nov; 7 Apr–7 May). **Habitat**: Forests, gardens. **Voice**: A sharp *jeet*. **Food**: Insects.

BLACKPOLL WARBLER

BIJIRITA DE CABEZA NEGRA

Dendroica striata   Pl. 41

DESCRIPTION. 5.25" (13 cm). Winter plumage olive green above, streaked with black, with two white wing bars. Throat and upper belly yellow with faint streaks. Lower belly and long undertail coverts white; soles of feet, and sometimes legs, yellow. Breeding male olive gray above, heavily streaked with black, with a black cap and whisker stripe, and broadly white cheeks. White below, with sides streaked with black. Legs and soles of feet yellow. Breeding female similar to nonbreeders but whitish below and heavily streaked; legs yellowish. **SIMILAR SPECIES.** (1) Black-and-white Warbler is black above, with narrow white stripes on head and back. (2) Winter Bay-breasted Warbler is yellowish green above with bolder wing bars, and has buff undertail coverts and usually unstreaked underparts. The short undertail coverts make tail appear long. Legs and feet are gray. (3) Pine Warbler is unstreaked above and its short undertail coverts make tail appear longer. Legs and feet are black. (4) Chestnut-sided Warbler is bright yellowish green above, white below. **RANGE.** Breeds in northern North America. Winters mainly in western Amazonia. **Status**: Common transient in Cuba and larger cays; mostly in central and eastern areas during fall, widespread in spring (17 Aug–10 Dec; Apr–7 Jun). **Habitat**: Open forests, coastal vegetation, thickets, often on or near the ground or in low levels in bushes or trees. **Voice**: Call a distinctive *zheeet*. Song a thin *zeet-zeet-zeet*, etc., on one high pitch. **Food**: Insects, fruits.

CERULEAN WARBLER

BIJIRITA AZULOSA

Dendroica cerulea   Pl. 42

DESCRIPTION. 4.5" (11 cm). *Male*: blue above, streaked with black on back, with two distinct white wing bars. White below with narrow black band on upper breast and black streaks on sides. *Female*: bluish green above; yellowish white below. Creamy superciliary broadening behind eye; sides faintly streaked. Back unmarked. Immature resembles adult female but is yellower with faintly blue gray crown.

SIMILAR SPECIES. Male Black-throated Blue Warbler has black throat and breast. Both sexes have a conspicuous small white spot on wing at base of primaries.

RANGE. Breeds in southeastern Canada and eastern United States, wintering from Colombia to Peru and Bolivia. **Status**: Very rare transient in Cuba (4 Aug–18 Oct; Apr). **Habitat**: Semideciduous woods, open forests, other groups of trees. **Voice**: A thin *chip*. **Food**: Insects.

BLACK-AND-WHITE WARBLER

BIJIRITA TREPADORA; BIJIRITA BLANCA Y NEGRA; RAYAPALO

Mniotilta varia   Pl. 40

DESCRIPTION. 5.25" (13 cm). In winter plumage, the upperparts and sides are streaked black and white; cheek patch is gray to black, more contrasted in male; underparts are white to buff in female. Cheek tinged with buff in female. In breeding plumage, male has black throat and cheeks; female has white chin and throat, and gray cheeks. Immature is similar to winter female, with white throat and gray cheeks, the immature female having a buff wash below. Bill large and slightly decurved. Habitually creeps along branches, pausing to peer under them. Often flocks with Oriente Warbler in eastern Cuba.

SIMILAR SPECIES. Male Blackpoll Warbler in breeding plumage has black cap, white cheek, and yellow legs.

RANGE. Breeds in Canada and eastern United States, wintering to Peru and West Indies. **Status**: Common winter resident and transient in Cuba, Isla de Pinos, and the cays (29 Jul–25 May). **Habitat**: Forests, gardens. **Voice**: A buzzy *chit*, not often heard but very like call of Yellow-headed and Oriente Warblers. **Food**: Mostly insects caught on bark of trunks and branches.

AMERICAN REDSTART

CANDELITA

Setophaga ruticilla   Pl. 44

DESCRIPTION. 5.25" (13 cm). *Male*: black, with bright orange patches on side, wing, and base of tail. Belly white. *Female*: Olive gray above; white below. Side, wing, and tail patches yellow. Immature similar to female, the male having yellow orange side patch, and often some black flecks on throat and chest. Very active and agile, characteristically dropping short distances through the branches in pursuit of flying insects. Constantly droops wings and spreads tail, showing orange or yellow markings.

SIMILAR SPECIES. None.

RANGE. Breeds in North America, wintering south to Ecuador, northwestern Brazil, and West Indies. **Status**: Common winter resident and transient in Cuba, Isla de Pinos, and many cays, including the smaller ones (6 Aug–24 May). Some birds remain in Cuba during summer, and two nests have recently been reported in Camagüey and La Habana. **Habitat**: Forests, gardens, swamps, rocky and mangrove-covered cays. **Voice**: Call is a loud *tschip*, more liquid than call of Northern Parula. Also a *srelee*. The song, commonly heard in Cuba, is a quick series of five or six notes, usually with a slurred conclusion. **Food**: Mostly insects caught on the wing, from leaves or in the air.

PROTHONOTARY WARBLER

BIJIRITA PROTONOTARIA

Protonotaria citrea Pl. 40

DESCRIPTION. 5.5" (14 cm). *Male*: intensely orange yellow head and underparts, with olive back and long black bill. *Female*: slightly duller. Both sexes have unmarked bluish gray wings and short, bluish gray tail with large white spots. Lower belly and undertail coverts white. Immature is similar to adult female with brownish gray wings, the female having a more olivaceous head that gradually blends with the color of the back.

SIMILAR SPECIES. (1) Blue-winged Warbler has black eye line, and wing bars. (2) Female Yellow Warbler has yellowish wing bars and tail spots.

RANGE. Breeds in eastern North America, wintering from Yucatán peninsula to Netherland Antilles and northern South America. **Status**: Uncommon transient in Cuba and Isla de Pinos (3 Aug–1 Nov; 28 Feb–15 Apr). **Habitat**: Forests, gardens, mangroves, forages near streams at medium height. **Voice**: A metallic *tink*, somewhat like notes of the waterthrushes. **Food**: Insects.

WORM-EATING WARBLER

BIJIRITA GUSANERA

Helmitheros vermivorus Pl. 40

DESCRIPTION. 5.25" (13 cm). Dark olivaceous brown above, with bright tan head and underparts, the crown has four bold black stripes. Immature is very similar with paler upper breast and tertials with rusty tips (usually visible only in the hand). Rummages noisily for insects among hanging dry leaves rather high in trees. Sexes alike.

SIMILAR SPECIES. Swainson's Warbler has dark brown crown and long creamy eyebrow, and is normally found on or near the ground.

RANGE. Breeds in eastern North America, wintering from Veracruz, Mexico to Panama. Bahamas and Greater Antilles. **Status**: Common winter resident and transient in Cuba, Isla de Pinos, and some larger cays (18 Aug–5 May). **Habitat**: Shady forests, thickets. **Voice**: Call is a dry, sharp *chip*. Song, rarely heard, is an insect-like buzz, *dzit-dzit*. **Food**: Insects.

SWAINSON'S WARBLER
BIJIRITA DE SWAINSON

Limnothlypis swainsonii   Pl. 40

DESCRIPTION. 5.5" (14 cm). Dark olivaceous brown above without wing bars; buff white to yellowish white below, washed with olive gray on sides. Long creamy eyebrow contrasts sharply with dark brown crown. Bill strikingly long. Sexes alike. A ground dweller, easily overlooked; frequently associates with Ovenbirds.
SIMILAR SPECIES. Worm-eating Warbler has bold black stripes on head and usually feeds rather high in trees.
RANGE. Breeds in southeastern United States, wintering in Yucatán peninsula and Belize, northern Bahamas, Cuba, Cayman Islands, and Jamaica. **Status**: Uncommon winter resident and transient in Cuba, Isla de Pinos, and some cays (15 Sep–16 Apr). **Habitat**: Humid shady areas in forests, usually with a dense understory. **Voice**: A short, metallic, *ziip* often difficult to locate, also a *sreee* flight call and a liquid *chip* not unlike that of American Redstart. **Food**: Insects.

OVENBIRD
SEÑORITA DE MONTE

Seiurus aurocapillus   Pl. 43

DESCRIPTION. 6"(15 cm). Olive above, with orange crown bordered by black stripes. Eye appears large, with a distinct white ring. White below, with throat and breast boldly streaked with black. Legs pink. Immature similar to adults, but with rusty-edged tertials (usually visible only in the hand). Feeds on the ground by flipping over leaves; walks with tail cocked.
SIMILAR SPECIES. Both (1) Northern Waterthrush and (2) Louisiana Waterthrush have conspicuous white eyebrow, unmarked brown crown, lack eye ring, and wag their tails. (3) Thrushes are larger, with unmarked crowns.
RANGE. Breeds in central and eastern North America, wintering to northern South America and West Indies. **Status**: Common winter resident and transient in Cuba, Isla de Pinos, and the larger cays (15 Aug–24 May). **Habitat**: Forests, typically with abundant leaf litter. **Voice**: A loud, sharp *chek*. Song a vigorous *teacher-teacher-teacher*. **Food**: Insects, fruit.

NORTHERN WATERTHRUSH
SEÑORITA DE MANGLAR

Seiurus noveboracensis   Pl. 43

DESCRIPTION. 6" (15 cm). Dark brown above, with whitish or yellowish eyebrow. Pale yellow or whitish below, with throat, breast, and sides streaked with brown. Immature has rusty-edged tertials (usually visible only in the hand). Feeds on the ground; pumps rump and tail frequently and rapidly, more commonly downward.

SIMILAR SPECIES. (1) Louisiana Waterthrush has somewhat longer bill, conspicuous white eyebrow broadening rearward (or behind the eye), and contrastingly buffy flanks. (2) Ovenbird has orange crown bordered with black stripes, and distinct eye ring.

RANGE. Breeds mostly in Canada, Alaska, and northern United States, wintering to northern South America. West Indies. **Status**: Common winter resident and transient in Cuba, Isla de Pinos, and many cays (14 Jul–28 May). **Habitat**: Swamps, lagoons, mangroves. **Voice**: A loud, metallic *chink*. Song is loud with accelerating staccato ending, *sweet sweet sweet swee-wee-wee chew chew chew.* **Food**: Mostly aquatic insects.

LOUISIANA WATERTHRUSH

SEÑORITA DE RÍO

Seiurus motacilla   Pl. 43

DESCRIPTION. 6″ (15 cm). Dark brown above, with conspicuous white eyebrow, flared and broadening behind the eye. Bill long; throat usually unmarked. White below, streaked with brown on chest and sides; flanks contrastingly buffy. Immature has rusty-tipped tertials (usually visible only in the hand). Feeds on the ground; pumps rear body upward or downward, rather slowly; also swings the tail sideways.

SIMILAR SPECIES. (1) Northern Waterthrush has somewhat shorter bill, uniformly yellow or yellowish white underparts, and eyebrow narrowing behind the eye; pumps rump and tail faster, usually downward. (2) Ovenbird has orange crown bordered with black stripes, and a distinct eye ring.

RANGE. Breeds in eastern North America, wintering from Mexico to Panama, rarely to northern South America; West Indies. **Status**: Common winter resident and transient in Cuba and Isla de Pinos (14 Jul–22 Apr). **Habitat**: Shady, stony rivers and creeks. **Voice**: A metallic *chink*, very similar to call of Northern Waterthrush but slightly softer. Song a loud *see-you see-you see-you chew chew to-wee.* **Food**: Mostly aquatic insects.

KENTUCKY WARBLER

BIJIRITA DE KENTUCKY

Oporornis formosus   Pl. 43

DESCRIPTION. 5.5″ (14 cm). Plain, bright olive above. Crown, cheeks, and sides of neck black; bold spectacle and entire underparts bright yellow. Female has slightly less extensive black markings. Immature similar to female but black areas washed with olive.

SIMILAR SPECIES. (1) Immature male Common Yellowthroat lacks yellow spectacle, and never appears solid bright yellow below. (2) Canada Warbler has blue gray upperparts, breast streaked with gray or black, and white undertail coverts. (3) Hooded Warbler has large yellow mask and spreads tail often, showing white spots.

RANGE. Breeds in eastern United States, wintering to Panama, rarely to northern South America. **Status**: Rare transient and very rare winter resident in Cuba, Isla de Pinos, and some larger cays (3 Aug–15 Apr). **Habitat**: Forest, thickets. **Voice**: A sharp *check*. **Food**: Insects.

MOURNING WARBLER

BIJIRITA DE CABEZA GRIS

Oporornis philadelphia   Pl. 43

DESCRIPTION. 5.25 " (13 cm). *Male*: plain olive above, with gray hood that is broadly black along its lower edge; underparts otherwise wholly bright yellow. *Female*: similar, with pale broken eye ring and pale gray or brownish gray hood without black. Immature like female; throat is yellower, and hooded effect is nonetheless dimly visible. All plumages have the pale pink feet characteristic of the genus.

SIMILAR SPECIES. (1) Female Common Yellowthroat is somewhat brownish yellow rather than pure yellow below, and lacks hooded or bibbed appearance. (2) Nashville Warbler has a small white area on lower belly, and dark feet.

RANGE. Breeds in northern North America, wintering from southern Nicaragua to northern South America. **Status**: Vagrant. Two sight records: Oct, 1 Mar (1987). La Habana and Holguín provinces. **Habitat**: Forest undergrowth, thickets, shrubs. **Voice**: A very distinctive metallic *jink*. **Food**: Insects.

COMMON YELLOWTHROAT

CARETICA

Geothlypis trichas   Pl. 43

DESCRIPTION. 5 " (13 cm). *Male*: plain olive green above, with a black mask. Yellow below, with brownish white belly and brownish sides. *Female*: similar, without mask and with thin whitish eye ring. Immature similar to female, but duller, often with brownish underparts, and male usually has traces of black on face. All plumages have short, plain wings and unmarked, slightly rounded tails.

SIMILAR SPECIES. (1) Kentucky Warbler is entirely bright yellow below, with yellow spectacles. (2) Female Mourning Warbler has gray or buff hood, and yellow belly.

RANGE. Breeds in North America and north of Mexico, wintering to Panama. Bahamas, Greater Antilles, and Dominica (Lesser Antilles). **Status**: Common winter resident and transient in Cuba, Isla de Pinos, and many cays (3 Sep–14 May). **Habitat**: Bushes, gardens, sawgrass savanna, marshes, understory in open forest. **Voice**: A loud, husky *tchep*. **Food**: Insects.

YELLOW-HEADED WARBLER
CHILLINA
Teretistris fernandinae Pl. 43

DESCRIPTION. 5.25 " (13 cm). Yellow head; plain gray back, wings, and tail; grayish white underparts. Bill slightly decurved. Foraging habits are highly variable. Feeds in flocks, among branches and foliage, on bark, and on the ground.

SIMILAR SPECIES. Oriente Warbler is all-gray above, extensively yellow below. Range does not overlap with Yellow-headed Warbler.

RANGE. Endemic to Cuba. **Status**: Common but restricted to central and western Cuba and Isla de Pinos, and also occurring on Cayo Cantiles. **Habitat**: Forests and thickets from sea level to high elevations. **Nesting**: Mar–Jul. Builds cup-shaped nest, made of grasses, rootlets, and other plant fibers. Lays two or three pale-green eggs, spotted with lilac and reddish brown at large end. **Voice**: A very noisy, rasping *shhh-shhh-shhh-shhh shhh-shhh*. **Food**: Insects, small lizards, spiders.

ORIENTE WARBLER
PECHERO
Teretistris fornsi Pl. 44

DESCRIPTION. 5.25 " (13 cm). Upperparts plain gray; face and underparts yellow; lower belly grayish. Bill slightly decurved. Like Yellow-headed

Warbler, has highly variable foraging habits. Forms small flocks outside breeding season.

SIMILAR SPECIES. (1) Female Prothonotary Warbler is olivaceous above, with white spots on tail. (2) Yellow-headed Warbler has entirely yellow head, and grayish chest and belly. (3) Female Canada Warbler has faint streaks on breast, and yellowish lores and eye ring.

RANGE. Endemic to Cuba. **Status**: Common but restricted to central and eastern Cuba and the larger cays north of Camagüey and Ciego de Ávila provinces. **Habitat**: Forests and thickets, from sea level to highest elevations of Sierra Maestra. **Nesting**: Mar–Jul. Builds cup-shaped nest, made of grasses, rootlets, and other plant fibers. Lays two or three very pale-green eggs, spotted with lilac and reddish brown at large end. **Voice**: A noisy and rasping *shhh-shhh-shhh-shhh-shhh-shhh* not quite as harsh as that of Yellow-headed Warbler. **Food**: Insects, spiders, small lizards.

HOODED WARBLER

MONJITA

Wilsonia citrina   Pl. 44

DESCRIPTION. 5.25" (13 cm). *Male*: plain olive back and wings, with striking black hood and yellow mask. Yellow below. Tail spotted with white. *Female*: similar; the male's face pattern more or less plainly apparent. Immature like female but duller. Forages from ground, where it hops about rapidly, to middle levels, often spreading the tail and flashing the white spots.

SIMILAR SPECIES. (1) Male Bachman's Warbler lacks masked appearance. Female has gray head. Both sexes have rather long, decurved bill. (2) Female Wilson's Warbler has olivaceous sides of head, blending inconspicuously with olive back; lacks tail spots. (3) Female Yellow Warbler has yellow inner webs of tail feathers, yellow edges to primaries, and faintly to boldly streaked underparts.

RANGE. Breeds in eastern North America, wintering to Panama. **Status**: Common transient and rare winter resident in Cuba, Isla de Pinos, and some larger cays (3 Aug–28 Apr). **Habitat**: Shady forests. **Voice**: A metallic *chink*. **Food**: Insects.

WILSON'S WARBLER

BIJIRITA DE WILSON

Wilsonia pusilla   Pl. 44

DESCRIPTION. 4.75" (12 cm). *Male*: plain olive back, wings, and tail, with glossy black cap and yellow underparts. *Female*: similar, the crown olive or with some black. Like other members of the genus, habitually flicks tail upward, swiveling on perch.

SIMILAR SPECIES. (1) Female Hooded Warbler has large yellow patch on sides of face and flashy white spots on tail. (2) Female Bachman's Warbler has gray wash on head, white undertail coverts, and white spots on tail.

(3) Female Yellow Warbler has yellow primary edges and inner webs of tail feathers.

RANGE. Breeds in Canada and western United States, wintering south to Panama. **Status**: Very rare transient in Cuba (2 Sep–12 Nov; Feb–12 Apr); one winter record: 20 Jan. Some individuals probably winter in Cuba. **Habitat**: Forests and gardens, usually at lower levels. **Voice**: A husky *chuck*. **Food**: Insects.

CANADA WARBLER
BIJIRITA DEL CANADÁ
Wilsonia canadensis Pl. 44

DESCRIPTION. 5.25" (13 cm). *Male*: plain bluish gray back, wings, and tail; yellow underparts. Forehead black, contrasting with yellow lores and eye ring. Bold black streaks on breast. Undertail coverts white. *Female*: similar but duller; the chest streaks less distinct. Immature similar to female, with greenish forehead; chest streaks are sometimes nearly absent.

SIMILAR SPECIES. (1) Kentucky Warbler is olive rather than blue gray above, and entirely yellow below. (2) Female Hooded Warbler has a yellow mask, plain yellow underparts, and white-spotted tail. (3) Oriente Warbler is gray above and lacks streaks or strong facial pattern.

RANGE. Breeds in Canada and northeastern United States, wintering from Colombia south to Peru and northern Brazil. **Status**: Very rare transient in Cuba (22 Sep–11 Oct). **Habitat**: Open forests near swamps, gardens. **Voice**: An occasional low *tchup*. **Food**: Insects.

YELLOW-BREASTED CHAT
BIJIRITA GRANDE
Icteria virens Pl. 43

DESCRIPTION. 7.5" (19 cm). Very large for a warbler. Unmarked dark olive above with bold white spectacle, heavy bill, and rather long tail. Lores black, to gray in females, bordered with white. Throat and breast brilliant yellow; belly and undertail coverts white. Solitary and shy, remaining hidden in undergrowth.

SIMILAR SPECIES. None.

RANGE. Breeds in southwestern Canada, United States, and Mexico, wintering to Panama. **Status**: Very rare transient in Cuba (24 Oct–22 Nov; 12 Feb–5 May). **Habitat**: Forest edges, dense thickets, tangles. **Voice**: A single *kook*. **Food**: Insects.

BANANAQUITS Coerebidae

Small, colorful, and arboreal birds with short tails and slender decurved bills, used to probe fruits and sip nectar from flowers. (**W**:1; **C**:1)

BANANAQUIT
REINITA

Coereba flaveola Pl. 44

DESCRIPTION. 4.5" (11 cm). Black above with bold white eyebrows and yellow rump. Whitish below, with yellow breast. A white spot on wing at base of primaries. Bill thin and decurved, with red spot on base. Immature is duller with yellowish eyebrow.

SIMILAR SPECIES. None.

RANGE. Mexico to northern Argentina; West Indies. **Status**: Rare visitor and possible permanent resident. Five specimens, one from Cayo Tío Pepe and four from the environs of Gibara. Eight recent records are from the cays north of Ciego de Ávila and Camagüey: three caught and banded and another sighted on Cayo Coco; three observed on Cayo Paredón Grande and another on Cayo Guillermo. Vulnerable. **Nesting**: Breeding has not been confirmed. **Habitat**: Coastal vegetation. **Voice**: A wheezing, energetic *tzeee-tzuiiitzeeetzeee*. Also a short *tsip*. **Food**: Nectar, insects, fruits.

TANAGERS Thraupidae

Small to medium-sized, colorful and arboreal birds with thick bills. A large subfamily, restricted to the New World. Most species are tropical; the few that breed in north temperate regions are migratory. Food is mainly fruits and insects. Males are much more colorful than females. (**W**:248; **C**:5)

RED-LEGGED HONEYCREEPER
APARECIDO DE SAN DIEGO; AZULITO

Cyanerpes cyaneus Pl. 44

DESCRIPTION. 5" (13 cm). *Male*: breeding plumage purplish blue. Wings, back, and tail are black; yellow wing linings are conspicuous in flight. Crown metallic turquoise blue. Legs and feet bright red. Nonbreeders are female like, with black wings and tail. *Female*: olive green above, with yellowish white underparts densely streaked with olive. Legs red. Bill long, slightly decurved. Found singly, in pairs, or in small flocks.

SIMILAR SPECIES. None.

RANGE. Mexico to southern Brazil. **Status**: Rare permanent resident in Cuba, but locally common in Sierra del Rosario and Sierra de los Órganos, Pinar del Río province, and high in Sierra Maestra, Granma province. Recently recorded from Cayo Coco. **Habitat**: Mountain and lowland forests, shrubs, and particularly where bottlebrush (*Callistemon citrinus*) trees are common. **Nesting**: Mar–Jul. Builds cup-shaped nest, made of grasses, rootlets, and other plant fibers. Lays two whitish eggs, with a greenish blue cast, spotted with reddish brown and lilac at the large end. **Voice**: Song begins with a paired *breep-breep* followed by long series of high-pitched notes *too-too-too-wee-wee-too-too-too*. Call, a repeated, short, high *tseep-tseep*; also *srelee* and a nasal *bizzj*. **Food**: Nectar, fruits, insects.

STRIPE-HEADED TANAGER

CABRERO

Spindalis zena Pl. 45

DESCRIPTION. 6" (15 cm). *Male*: gaudily marked with orange, chestnut, black and white. Head is black with very bold white stripes; wing has white patch at base of primaries and extensive white edgings. Throat and chest rich chestnut brown, with darker upper chest. Belly grayish. *Female*: olive gray above, with indistinct markings on head, whitish spot at base of primaries, and pale grayish brown underparts. Adult females often have orange flush on shoulders, rump, and upper breast. Juvenile similar to female. Bill short and heavy in both sexes.

SIMILAR SPECIES. None.

RANGE. Bahamas, Greater Antilles, Grand Cayman, and Cozumel. **Status**: Common permanent resident in Cuba, Isla de Pinos, and many of the larger cays off north and south coasts: Santa María, Coco, Cinco Leguas, Cantiles. **Habitat**: Open forests, thickets, mountain forests, from sea level to the highest altitudes. **Nesting**: Feb–Jul. Builds cup-shaped nest quite high in branches. Lays two or three white eggs, sparsely covered with brown and blackish spots. **Voice**: Song, a variable, jumbled series of high *tseeps* and buzzes, e.g., *weeze-weeze-weeze-tee-tee-bizz-weeze-weeze*. Most characteristic call is a thin, high-pitched *tsiiiiiii*; also a short, high *tseep*, very similar to Cape May Warbler. **Food**: Mainly fruits, small seeds, insects.

SUMMER TANAGER
CARDENAL; TANAGRA
Piranga rubra Pl. 45

DESCRIPTION. 7.5" (19 cm). *Male*: entirely bright red, with darker wings and tail. Immature male is olive above, with slightly darker wings and orange yellow underparts sometimes blotched with red. *Female*: similar to immature male. Both sexes have heavy and rather short bill, which is paler and somewhat longer than in other *Piranga* spp. Yellowish underwing coverts are visible in flight. Forages alone or in pairs.

SIMILAR SPECIES. Breeding male Scarlet Tanager has black wings and tail; winter male and female are greener, with darker wings and tail, and shorter bill.

RANGE. Breeds in southern North America, wintering to Bolivia and Brazil; Bahamas and Cuba. **Status**: Rare winter resident and transient in Cuba, Isla de Pinos, and some larger cays (1 Sep–28 Apr). **Habitat**: Forests, thickets, gardens. **Voice**: A staccato *pi-ki-tuck*; also a whistle, *tooee*. **Food**: Large insects, fruits, and seeds, most notably those of royal palm.

SCARLET TANAGER
CARDENAL ALINEGRO; TANAGRA; CARDENAL DE ALAS NEGRAS
Piranga olivacea Pl. 45

DESCRIPTION. 7" (18 cm). *Male*: winter plumage olive green above, with blackish wings and tail; dusky yellow below. Breeding plumage, seen only during northward migration in spring, is bright red, with black wings and tail. *Female*: similar to winter male, with greenish brown wings. Both sexes have yellowish, heavy, and rather short bill. Usually sits quietly on high branches, making quick rushes for food.

SIMILAR SPECIES. Immature male and female Summer Tanager have a slight orange wash below; wings and tail are more nearly the tone of the body; bill is longer.

RANGE. Breeds mostly in northeastern United States, wintering from Panama to Bolivia. **Status**: Rare transient in Cuba, Isla de Pinos, and some large northern cays, more common in fall than in spring (21 Sep–6 Nov; 16 Feb–18 Apr). **Habitat**: Open semideciduous forests, gardens. **Voice**: A burry, low *chip-burr* with a twanging undertone. **Food**: Large insects, fruits, seeds.

WESTERN TANAGER*

CARDENAL DEL OESTE

Piranga ludoviciana   Pl. 51

DESCRIPTION. 7" (18 cm). *Male*: winter plumage, greenish yellow head, of-ten faintly red. Middle back, wings, and tail are mostly black. Two rather pronounced yellow wing bars. Lower back and belly yellow. In breeding plumage, the head is red. *Female*: grayish back, with olive yellow hindneck and rump. Head is without red or occasionally with a few traces. Wing bars whitish or pale yellow. Underparts yellow. Bill is rather heavy as in other members of the genus. Immature resembles female.

SIMILAR SPECIES. Female orioles have sharply pointed bills.

RANGE. Western North America, wintering south to Central America. **Status**: One unconfirmed record in Cuba: 2 Jan (1978). Cárdenas. A recent sight record in La Habana. **Habitat**: Woodlands, open country. **Voice**: Call a slurred *pit-er-ick*. **Food**: Insects, fruits.

GRASSQUITS AND SPARROWS Emberizidae

Small or medium-sized birds with conical bills and more or less terrestrial habits. Most are cryptically colored. Food consists of seeds, fruit, and insects. (**W**:285; **C**:14)

GREEN-TAILED TOWHEE

GORRIÓN DE COLA VERDE

Pipilo chlorurus   Pl. 48

DESCRIPTION. 7.25" (18 cm). Olive above, with a gray head with reddish brown crown and black whisker. Gray below, with rather sharply defined white throat. Tail long and rounded; bill short and conical.

SIMILAR SPECIES. None.

RANGE. Breeds in western United States, wintering to Mexico. **Status**: Va-grant. One record: 8 Jan (1964). Trinidad. **Habitat**: Semiarid shrubbery. **Voice**: A soft catlike *mew*. **Food**: Insects, seeds.

CUBAN BULLFINCH

NEGRITO

Melopyrrha nigra   Pl. 47

DESCRIPTION. 5.5" (14 cm). Bill strikingly thick and black. *Male*: shiny black, with white patch on wing. *Female*: similar but duller, with smaller wing patch. Immature male resemble female, but with glossy black head and flight feathers. Juvenile similar to female, with greenish-tipped feath-ers. Often forages in small flocks.

SIMILAR SPECIES. None.

RANGE. Grand Cayman Island and Cuba. **Status**: Common permanent resident on Cuba, Isla de Pinos, and several larger cays of Archipiélago de Sabana-Camagüey, and Cayo Cantiles in Archipiélago de los Canarreos. **Habitat**: Thickets and forests, from sea level to moderate altitudes. **Nesting**: Mar–Aug. Nest is large and globular, with side entrance, made of dry grasses and leaves, rootlets, hair, and feathers. Lays three to five whitish eggs with a greenish cast, spotted with reddish brown and lilac concentrated at larger end. **Voice**: A thin, prolonged, and melodious warble, *ti-ti-tisissiiiitssiiiitsiiii-tooee-toeee*. Call a staccato *chi-dip*; also a thin *tsee*, often repeated. **Food**: Fruits, seeds, insects.

CUBAN GRASSQUIT
TOMEGUÍN DEL PINAR; SENSERENICO
Tiaris canora   Pl. 46

DESCRIPTION. 4.5″ (11 cm). *Male*: dark olive above with gray crown. Mask and large breast patch are black separated by a bold yellow-orange collar that continues back to a point behind the eye. Belly gray. *Female*: similar, with chestnut mask and paler yellow collar. Underparts gray. Juvenile similar to female but duller. Social outside the breeding season. When feeding in grass, these birds habitually jump to reach seeds higher on stems.

SIMILAR SPECIES. (1) Yellow-faced Grassquit has small throat patch and or-ange or yellow eyebrow. (2) Female Black-faced Grassquit has olivaceous gray face, throat, and breast.

RANGE. Endemic to Cuba. **Status**: Common in some regions, but declining near human settlements. Absent from Isla de Pinos and the cays. **Habitat**: Coastal shrubbery, semideciduous woods, pine forests, thickets near culti-vated fields. **Nesting**: Apr–Jun. Nest is large and globular with side en-trance, made of dry grasses, rootlets, hair, and other plant fibers. Lays two or three white eggs, with a greenish gray cast, spotted with brown and lilac concentrated at large end. **Voice**: Song, a rather harsh *chiri-wechee-wechee, chibiri-wechee*; also *tsit-tsit-tillio, tsit-tsit-tillio*. Call, a high, often repeated *tsit*. **Food**: Seeds, small fruits.

YELLOW-FACED GRASSQUIT
TOMEGUÍN DE LA TIERRA; VIUDITO
Tiaris olivacea Pl. 46

DESCRIPTION. 4.5″ (11 cm). *Male*: olive above; grayish below. Conspicuous orange eyebrow and throat patch. Size of black patch on breast increases with age. *Female*: similar, but pale yellow eyebrow and throat patch and no breast patch. Juvenile similar to female, with paler eyebrow and throat patch. Social outside the breeding season.

SIMILAR SPECIES. (1) Male Cuban Grassquit has large black mask and a bold yellow-orange collar. Female has chestnut mask and yellow collar. (2) Black-faced Grassquit lacks yellow or orange tones.

RANGE. Mexico to northern South America; Greater Antilles and Cayman Islands. **Status**: Common permanent resident in Cuba, Isla de Pinos, and some larger cays. **Habitat**: Thickets, shrubbery and agricultural land from sea level to moderate elevations. **Nesting**: Year-round. Nest is low, globu-lar, with side entrance, made of dry grasses, rootlets, hair, and other plant fibers. Lays two to four white eggs with a bluish cast, spotted with brown and lilac concentrated at large end. **Voice**: Song, a weak, pleasant trill, *tseee-tseee-tseee-tseee-tseee-tseee.* Call, a high, thin *tsit*. **Food**: Seeds, small fruits, tender shoots.

BLACK-FACED GRASSQUIT
TOMEGUÍN PRIETO; BARBITO
Tiaris bicolor   Pl. 46

DESCRIPTION. 4.5″ (11 cm). *Male*: dark olive back; sooty, lusterless black head and underparts. *Female*: plain olivaceous gray above; paler below. Immature duller. As in all members of this small genus, wing bars are entirely lacking. Usually feeds on the ground.

SIMILAR SPECIES. Females of both (1) Cuban and (2) Yellow-faced Grassquits have faded yellow patterns on head and throat.

RANGE. Caribbean Islands and northern South America. **Status**: Very rare permanent resident only on Cayo Tío Pepe, northeast of Isabela de Sagua. Current status of this population is unknown. Also recorded on Cayo Guillermo. Vulnerable. **Habitat**: Low thickets and shrubbery, coastal vegetation. **Nesting**: Apr–Jun. Nest is low and globular, with an entrance in the side or bottom. Usually lays three whitish eggs, heavily flecked at the broad end. **Voice**: Call note is a soft, musical *tsip*. Song is a buzzing *tik-tseee, tik-tik-tseee*. **Food**: Seeds, small fruits.

SAFFRON FINCH
GORRIÓN AZAFRÁN
Sicalis flaveola

Not Illustrated

DESCRIPTION. 5.5″ (14 cm). All yellow with orange crown in adult. Immature duller with grayish head, white underparts, and yellow breast.

SIMILAR SPECIES. Yellow Warbler usually has rusty streaks on underparts, thinner bill, and arboreal habits.

RANGE. South America; in West Indies, introduced to Jamaica and Puerto Rico. **Status**: Vagrant. One record: Oct (1996). Guantánamo province. **Habitat**: Open country near the coast. **Voice**: Call a sharp *pink*. **Food**: Seeds.

ZAPATA SPARROW
CABRERITO DE LA CIÉNAGA
Torreornis inexpectata   Pl. 47

DESCRIPTION. 6.5″ (17 cm). Grayish olive above with darker streaks and chestnut crown. White throat, bordered by black whisker; yellow breast and belly, grayish olive sides. Three races are known: Zapata and Cayo Coco races are quite similar, with rather bright colors. The easternmost race, from Guantánamo province, is duller, with the crown almost gray. Immature is duller than adult, being dark grayish olive above and light yellow below. A weak flier.

SIMILAR SPECIES. None.

RANGE. Endemic to Cuba. **Status**: Localized, vulnerable, and rare endemic resident with populations in three widely separated areas: Zapata peninsula, specifically north of Santo Tomás; Cayo Coco, in northern Ciego de Ávila province; and the coastal region between Baitiquirí and Imías in Guantánamo province. **Habitat**: Quite different in the three areas. Zapata: extensive sawgrass prairies that are flooded to a depth of 0.3–0.6 m for about half the year. Cayo Coco: dry and semi-wet forest and shrubbery, and to a lesser extent, coastal vegetation. Baitiquirí: thorn scrub, cacti and scattered trees. **Nesting**: Apr–Jun. Only three nests have been discovered, all in Zapata swamp, where they were found to be cup shaped and placed low on top of tussocks, surrounded by sawgrass. Eggs are white with a green wash and spotted brown. Two eggs comprise the largest clutch yet discovered. **Voice**: Song is a high-pitched, buzzy, and cascading *tzi-tzi-tzi-iiitziiii*. Call is a thin *tseep*; also a repeated *pit* and a repeated, mettalic *oing*. **Food**: At Zapata swamp during the flooded season, feeds extensively on eggs of water snails (*Pomacea*). Also small lizards. During the dry season, feeds on seeds and small invertebrates, often on the ground. At Cayo Coco, feeds mostly on insects, both on the ground and among bromeliads. Also seeds and fruits found in leaf litter. Eastern birds also feed on cactus fruits.

CHIPPING SPARROW
GORRIÓN DE CABEZA CARMELITA
Spizella passerina   Pl. 47

DESCRIPTION. 5.5" (14 cm). Winter plumage brown, narrowly streaked on back and finely on crown. Lower mandible pale; ear patch brownish. Rump unstreaked lead gray. An inconspicuous pale grayish brown eyebrow; eye line and lores blackish. Underparts gray, extending to hindneck. In breeding plumage, the crown is reddish brown; eyebrow conspicuously white; bill, eye line, and lores are dark; and underparts are pale gray. Both sexes have fairly long, notched tail and whitish wing bars. Immature similar to winter adult, but darker.

SIMILAR SPECIES. Clay-colored Sparrow has whitish coronal stripe, conspicuous dark-bordered brown ear patch, decidedly pale gray side of neck and nape, and brown rump.

RANGE. Breeds throughout North America and south to Nicaragua, wintering within this range. **Status**: Vagrant. Four records: One from Gundlach's collection (last century); 25 Nov (1991); 20 Nov (1996); Nov (undated). La Habana, Cayo Coco. **Habitat**: Bushy areas. **Voice**: A dry, sharp *tseep*. **Food**: Seeds, insects.

CLAY-COLORED SPARROW
GORRIÓN COLORADO
Spizella pallida   Pl. 47

DESCRIPTION. 5.5" (14 cm). Similar to Chipping Sparrow but in all plumages is paler and brighter with more pronounced contrasts in the face. Brown above, with striped head and streaked back. Whitish coronal stripe and eyebrow and dark whisker are conspicuous. Distinct brown ear patch bordered with dark brown. Underparts whitish. Nape and side of neck are pale gray; brown rump appears uniform with back, unlike the contrasting lead-gray rump of the Chipping Sparrow. Tail fairly long and notched. Immature similar to adult, with breast and buff sides.

SIMILAR SPECIES. Winter adult and immature Chipping Sparrow have lead-gray rump and lack both coronal stripe and dark forward border to the cheek patch.

RANGE. Breeds in northern North America, wintering to Mexico. **Status**: Rare transient and winter resident to Cuba (15 Oct–5 Dec; 26 Jan). **Habitat**: Bushy savanna. **Voice**: A thin *tseep*. **Food**: Seeds, insects.

LARK SPARROW
GORRIÓN DE UÑAS LARGAS

Chondestes grammacus Pl. 47

DESCRIPTION. 6.5″ (17 cm). Brown above with dark streaks, with bold chestnut and white stripes on head. Whitish below, with distinct black spot in the center of breast. Tail long and graduated, with white corners.

SIMILAR SPECIES. Adult White-crowned Sparrow has head boldly striped with black and white; tail and breast are unmarked.

RANGE. Breeds in southern Canada, United States, and Mexico; Canadian and most U.S. breeders winter in southern United States and Mexico. **Status**: Vagrant. Four records: 12 Dec; 10 Feb (1958); 25 Mar (1965); one undated from last century. La Habana. **Habitat**: Bushy savanna. **Voice**: A sharp *tink* or *jeeet*. **Food**: Insects, seeds.

SAVANNAH SPARROW
GORRIÓN DE SABANA

Passerculus sandwichensis Pl. 47

DESCRIPTION. 5.75″ (15 cm). Brown above, heavily streaked. Whitish below, with breast and sides streaked, the breast often having a central black spot. Head has pale coronal stripe, conspicuous yellowish or whitish lores and eyebrow and distinct whisker. Tail short and notched; legs and feet pink. Flies only short distances when flushed.

SIMILAR SPECIES. (1) Grasshopper Sparrow is slightly smaller and unmarked below. Outer tail feathers are sharply pointed. (2) Lincoln's Sparrow has gray eyebrow and is more warmly toned and finely streaked below.

RANGE. Breeds in North America and south to Guatemala, wintering to Honduras; Bahamas, Cuba, Cayman Islands, and Swan Islands. **Status**: Uncommon winter resident and transient on Cuba (11 Oct–25 Apr). **Habitat**: Grassy fields, savanna, pastures, low coastal vegetation. **Voice**: Song is sometimes heard: *tsip tsip tsip tseeee-tsee-ayyyy*. Most common call is a simple *tse* or thin *tsip*. **Food**: Seeds, fruits, insects.

GRASSHOPPER SPARROW
CHAMBERGUITO; CODORNIZ DE LA TIERRA

Ammodramus savannarum Pl. 47

DESCRIPTION. 5″ (13 cm). Appears flat-headed with a low forehead, proportionally large bill, and prominent eye ring. Brown above, streaked. Buff breast and sides. Streaked underparts; whitish belly. Pale coronal stripe is bordered by dark lateral stripes. Most birds have a fairly conspicuous orange mark forward of the eye and yellow at the bend of wing. Tail is short and feathers are sharply pointed. Flies only short distances, quickly hiding among grasses.

SIMILAR SPECIES. (1) Savannah Sparrow has heavily streaked breast and sides, a conspicuous whisker, and longer tail. (2) Lincoln's Sparrow is streaked below, with bold gray eyebrow. Both species lack sharply pointed tail feathers.

RANGE. Breeds from southern Canada to Panama, at isolated localities in northwestern South America, and locally in Jamaica, Hispaniola, and Puerto Rico. **Status**: Uncommon and inconspicuous winter resident in Cuba. Transient in the cays (12 Oct–1 May). **Habitat**: Grassy fields, savanna, rice plantations. **Voice**: A faint insect-like *tic-zzzzzzzzzz*; also a more prolonged jumble of rapid high notes. **Food**: Seeds, insects.

LINCOLN'S SPARROW
GORRIÓN DE LINCOLN
Melospiza lincolnii   Pl. 47

DESCRIPTION. 5.5" (14 cm). Brown above, with streaks. Crown has medium gray stripe bordered by reddish brown stripes. Eye ring white and fairly prominent. Particularly characteristic is a warm buff malar stripe with its forward edge neatly outlined by a dark whisker. Throat and belly are white, with buff breast band and sides. Throat, breast, and sides are narrowly streaked. A distinct dark brown chest spot may be present. Furtive and difficult to observe.

SIMILAR SPECIES. (1) Grasshopper Sparrow is mostly unmarked below; tail feathers are pointed. (2) Savannah Sparrow has heavily and more coarsely streaked breast and sides, and lacks warm buffy undertone across breast.

RANGE. Breeds in northern and western North America, wintering to Honduras, casually to Cuba, Jamaica, and Puerto Rico. **Status**: Very rare transient and winter resident in Cuba (12 Oct–28 Nov; 8 Jan–Mar). **Habitat**: Shrubby savanna, forest undergrowth, thickets, gardens. **Voice**: A hard *chup* and a peculiar buzzy *dzzzzzz*. **Food**: Seeds, insects.

WHITE-CROWNED SPARROW
GORRIÓN DE CORONILLA BLANCA
Zonotrichia leucophrys   Pl. 47

DESCRIPTION. 7" (18 cm). Back, wings, and tail brown. Crown boldly striped in black and white; face and neck gray; throat whitish; belly mostly gray. Bill is orange or pink, with dusky tip. Two narrow but distinct wing bars. Immature has head striped with buff and dark brown.

SIMILAR SPECIES. (1) Lark Sparrow has chestnut and white head stripes and white tail corners. (2) Female House Sparrow resembles immature but is dingier and lacks the bold head stripes. Single wing bar is indistinct. (3) Bobolink has streaked sides, unmarked wing, and sharply pointed tail feathers.

RANGE. Breeds in northern Canada and western United States, wintering across southern United States south to Mexico, Bahamas, and Jamaica.

Status: Rare transient in Cuba, and a very rare winter resident (2 Oct–22 Mar). **Habitat**: Open forests, grassy fields. **Voice**: A thin *seeet*. **Food**: Seeds, insects.

GROSBEAKS AND BUNTINGS Cardinalidae

Small to medium-sized birds with stout conical bills. This group is represented in Cuba by six species occurring as migrants or winter residents from North America. Males of these are notably colorful; females are much less so. They are seed, fruit, and insect eaters, and are found in light forest or shrubbery. (**W**:47; **C**:6)

ROSE-BREASTED GROSBEAK
DEGOLLADO

Pheucticus ludovicianus Pl. 48

DESCRIPTION. 7.75" (20 cm). *Male*: in winter plumage, the head and back are very dark brown or black; white spots on wing coverts and primaries; and large white patches on outer tail feathers. Wing linings pink; rump white with gray barring. Red triangular patch on upper breast tapers into white lower breast and belly. Sides are streaked. Breeding plumage similar but head and back are black, and sides and rump are entirely white. *Female*: brown above, streaked; whitish stripes on head; whitish wing bars. Wing lining yellow. Pale gray below, streaked. Both sexes have very heavy pale bill. Immatures of both sexes are similar to female, but male has pink wash on upper breast.

SIMILAR SPECIES. None.

RANGE. Breeds in eastern and north central North America, wintering from middle Mexico south to Peru. **Status**: A rare transient and very rare winter resident in Cuba, Isla de Pinos, and some larger cays (23 Sep–7 May). **Habitat**: Forests, gardens. **Voice**: A strong, harsh *eek*. **Food**: Seeds, fruits.

BLUE GROSBEAK
AZULEJÓN

Guiraca caerulea Pl. 48

DESCRIPTION. 6.75" (17 cm). *Male*: blue, with reddish brown bars on dull black wing. *Female*: brown above, with dark wings and tail. Underparts are paler, with whitish throat. Rump and shoulder tinged with blue. Wing bars pale brown. Bill heavy. Immature is more uniformly and more richly orange brown than adult female. Habitually flicks tail.

SIMILAR SPECIES. (1) Indigo Bunting is smaller, with smaller bill and without conspicuous wing bars. (2) Female Lazuli Bunting is smaller with pale blue rump and less distinct white wing bars. In general, Blue Grosbeak always appears larger and more pointy headed than the *Passerina* buntings.

RANGE. Breeds from United States to Costa Rica, wintering to Panama. **Status**: Common transient and very rare winter resident in Cuba, Isla de Pinos, and some larger cays (1 Sep–2 May). **Habitat**: Forests, coastal vegetation, gardens. **Voice**: A sharp *chink*. **Food**: Seeds, fruits.

LAZULI BUNTING
MARIPOSA AZUL
Passerina amoena   Pl. 46

DESCRIPTION. 5.5" (14 cm). *Male*: winter plumage turquoise blue above, edged with brown on head and rump; wings blackish with two white bars; tail edged with blue. A broad cinnamon band on breast and sides. Belly whitish. In breeding plumage, head and rump are pure turquoise. *Female*: brown above, with bluish rump and two buff wing bars. Wings and tail are dark brown with a trace of blue. Unmarked below, with a buff wash on throat and breast, and whitish belly. Immature similar to female, with faint streaks on breast and brown rump.
SIMILAR SPECIES. (1) Female Indigo Bunting has dim streaks on breast, obscure or no wing bars, and brown rump. Breeding male is entirely blue. (2) Male Blue Grosbeak is larger, much deeper blue, and has two distinct reddish brown wing bars.
RANGE. Breeds in central and western North America, wintering to Mexico. **Status**: Vagrant. One record: 29 Mar (1960). La Habana province. **Habitat**: Bushy areas. **Voice**: A hard *spit*. **Food**: Insects, seeds.

INDIGO BUNTING
AZULEJO
Passerina cyanea   Pl. 46

DESCRIPTION. 5.5" (14 cm). *Male*: winter plumage brown above with traces of blue above and below, strongest on rump. Breeding plumage is indigo blue. *Female*: brown above; paler below, with lightly streaked breast and faint brown wing bars. Immature similar to female. Bill small and conical. Forms small flocks.
SIMILAR SPECIES. (1) Blue Grosbeak is larger with conspicuous rich-brown wing bars. (2) Female Lazuli Bunting is unmarked below, with bluish rump; male has cinnamon band on chest.
RANGE. Breeds in southwestern, central, and eastern North America, wintering to Panama and rarely to northern South America; Bahamas and Greater Antilles. **Status**: Common winter resident and transient in Cuba, Isla de Pinos, and some larger cays (21 Sep–13 May). **Habitat**: Open forests, stands of exotic trees such as teak and *Casuarina* spp., bushy areas. Often feeds on ground. **Voice**: A snappy characteristic *spit* and a ringing *zheeeg*; also a short *biiz*. **Food**: Seeds, fruits, insects.

PAINTED BUNTING
MARIPOSA; VERDÓN

Passerina ciris   Pl. 46

DESCRIPTION. 5.5" (14 cm). *Male*: astonishingly colorful, with purple blue head, yellow green back, and bright red rump and underparts. *Female*: very plain, bright olive green above and olive-yellow below, with faint eye ring. Immature male similar to adult female but brighter green, particularly on head.

SIMILAR SPECIES. (1) Female Black-faced Grassquit is smaller, duller, and more olivaceous green, with darker bill. (2) Lazuli Bunting lacks red or green tones, and has wing bars.

RANGE. South central and southeastern United States and northeastern Mexico, wintering to Panama, Bahamas, Cuba, and Jamaica. **Status**: Common transient and a rare winter resident on Cuba, Isla de Pinos, and some large northern cays (15 Oct–29 Apr). **Habitat**: Thickets, shrubbery. **Voice**: A sharp, liquid *chip*. **Food**: Seeds, small fruits.

DICKCISSEL
GORRIÓN DE PECHO AMARILLO

Spiza americana   Pl. 48

DESCRIPTION. 6.25" (16 cm). *Male*: brown with black streaks above and a chestnut shoulder patch. Head is gray with yellow eyebrow. White chin, and black triangular throat patch. Breast yellow; belly pale gray. *Female*: similar, without black throat patch, and with less extensively yellow underparts and less conspicuous shoulder patch. Immature similar to female, but considerably duller with faintly streaked flanks.

SIMILAR SPECIES. Female House Sparrow is unstreaked grayish brown below, with broad eyebrow and obvious wing bars.

RANGE. Breeds in central North America, wintering to northern South America. **Status**: Very rare transient in Cuba (3 Sep–23 Nov; 18 Mar–2 May). **Habitat**: Open areas, shrubbery and thickets, cultivated fields. **Voice**: A short, very distinctive *bzzt* with the quality of an electric buzzer.

BOBOLINKS, BLACKBIRDS, MEADOWLARKS, GRACKLES, AND COWBIRDS Icteridae

Medium-sized and generally glossy black birds. Some species are strikingly patterned in yellow or orange. Bill is conical, long, and pointed. Food is varied, from insects and small vertebrates to fruits and nectar. Some species forage extensively on the ground, whereas others are strictly arboreal. Many species are notably noisy but the group contains some fine singers, particularly among the orioles. Highly gregarious, some blackbirds gather in huge roosts. Most North American species are migratory and sexually dimorphic. (**W**:98; **C**:14)

BOBOLINK
CHAMBERGO
Dolichonyx oryzivorus   Pl. 45

DESCRIPTION. 7" (18 cm). Winter plumage warm buffy brown, streaked above with dark brown; head has two dark stripes. Underparts pale buffy brown; sides lightly streaked with dark brown. *Male*: breeding plumage black, with large yellowish-buff patch on hindneck and white scapulars and lower back. Black plumage is fringed with buff in early spring, the tips gradually wearing off as the season advances. *Female*: breeding plumage similar to winter plumage but somewhat paler. Juvenile similar to winter female but with more buff above, unmarked and yellower below. Bill short, wide at base. Tail feathers sharply pointed in all plumages.

SIMILAR SPECIES. (1) Savannah Sparrow has heavily streaked chest and like the following two species, has normal unmodified tail feathers. (2) Female House Sparrow has unmarked crown and underparts. (3) Immature White-crowned Sparrow has white wing bars.

RANGE. Breeds in central North America, wintering in South America from Peru to Argentina. **Status**: Uncommon transient (13 Aug–12 Dec; 5 Mar–12 Jun). **Habitat**: Coastal vegetation, rice fields, savanna. **Voice**: In flight, a soft and very distinctive *pink*. **Food**: Seeds, insects.

RED-SHOULDERED BLACKBIRD
MAYITO DE CIÉNAGA; CHIRRIADOR
Agelaius assimilis   Pl. 49

DESCRIPTION. 8.25" (21 cm). *Male*: black with an orange red patch on shoulder. *Female*: entirely black. Both sexes sing. First-year male similar to female, with brownish patch on shoulder. Both sexes have rather square tail with slightly pointed feather tips. Juvenile dull black. Very social, forming flocks with Cuban Blackbird and Tawny-shouldered Blackbird in the nonbreeding season.

SIMILAR SPECIES. (1) Male Tawny-shouldered Blackbird is smaller, has orange buff shoulder patch, smaller bill, and notched, rounded tail. (2) Cuban Blackbird is slightly larger with larger bill and round-tipped tail feathers.

RANGE:. Endemic to Cuba. **Status**: Rare but locally common in western and central Cuba. **Habitat**: Swamps, rice fields, marshes. **Nesting**: May–Aug. Builds a deep cup-shaped nest among tall sawgrass and reeds and lays three or four pale bluish-white eggs, heavily spotted and scrawled with reddish brown and pale purple. **Voice**: Frequently repeated loud harsh creaks (hence *Chirriador*, which means creaker), *o-weeheeee-o-wee-heeee*. Also a short *chek-chek*. **Food**: Insects, seeds, nectar.

TAWNY-SHOULDERED BLACKBIRD
MAYITO; TOTÍ MAYITO
Agelaius humeralis   Pl. 49

DESCRIPTION. 8" (20 cm). Black; both sexes with an orange buff shoulder patch, smaller and duller in female. Tail feathers rounded at tip. Juvenile is dull black, with smaller and paler shoulder patch, sometimes visible only in flight. When giving the *cheek* call, birds invariably cock the tail. Highly social.

SIMILAR SPECIES. Male Red-shouldered Blackbird is larger and has orange red shoulder patch. Female is entirely black, with larger bill. Both sexes have slightly pointed tail feathers.

RANGE. Cuba and Hispaniola. **Status**: Common permanent resident on Cuba and several larger cays on both coasts. Not known from Isla de Pinos. **Habitat**: Open forests, forest edge, agricultural fields, pastures, swamp edges, coastal thickets. **Nesting**: Apr–Aug. Builds cup-shaped nest in a tree, of grasses, hair, and occasionally some feathers. Generally lays three or four greenish-white, brown-spotted eggs. **Voice**: Short whistles, harsh creaks, *weee-weee*, and a nasal *enk*. Also a strong *cheek*, often re-peated. **Food**: Bees, seeds, nectar, fruits, small lizards.

EASTERN MEADOWLARK
SABANERO
Sturnella magna   Pl. 48

DESCRIPTION. 9″ (23 cm). Brown above, streaked with dark brown; flat head with bold brown and buff stripes. Bill long and straight. Bright yellow below, with striking black V-shaped band across breast and whitish belly. Sides streaked with dark brown. In winter, black breast band scaled with yellow. Juvenile similar to adult but paler with gray throat. Tail is short and very conspicuously and extensivly edged with white. Mostly terrestrial, occurring singly or in small groups. Flight is low, with rapid wingbeats and short glides. Often flicks tail when on the ground.

SIMILAR SPECIES. None.

RANGE. Breeds throughout eastern and southern North America and south to northern Brazil. **Status**: Common permanent resident in Cuba, Isla de Pinos, and Cayo Romano. Single report from Cayo Coco. **Habitat**: Savanna, pastures, open forests with large grassy glades, dry swamps. **Nesting**: Apr–Jul. Builds domed nest among grasses. Lays four or five white eggs, with reddish brown spots concentrated toward the large end. **Voice**: A sweet, slurred whistle: *tsee-ya, tsee-yair*. Also a bubbling chatter. **Food**: Insects, small lizards, worms, seeds.

YELLOW-HEADED BLACKBIRD
MAYITO DE CABEZA AMARILLA
Xanthocephalus xanthocephalus   Pl. 49

DESCRIPTION. 9.5″ (24 cm). *Male*: black with orange yellow head and breast, and a white wing patch. *Female*: smaller, dusky brown with dull yellow face, throat, and breast. Lower breast streaked with white. Wing patch absent. Immature similar to female, with white wing patch barely suggested in immature male.

Similar species. None.

Range. Breeds in western North America, wintering to Mexico. **Status**: Vagrant. Four records: one last century; fall (two); 10 Dec. **Habitat**: Farmland, fields. **Voice**: A sharp *ktuck*. **Food**: Seeds, fruits, large insects.

CUBAN BLACKBIRD
TOTÍ; CHONCHOLÍ

Dives atroviolacea   Pl. 49

Description. 10.5" (27 cm). Entirely black, with violaceous metallic sheen above. Brown eye. Female slightly smaller. Juvenile dull black.

Similar species. (1) Adult Greater Antillean Grackle has yellow eye, longer, thinner bill, and creased or keel-shaped tail. (2) Female Redwinged Blackbird is smaller, with slightly pointed tips to the tail feathers.

Range. Endemic to Cuba. **Status**: Common, although absent from Isla de Pinos and offshore cays. **Habitat**: Montane forests, farmland, parks, villages. **Nesting**: Mar–Jul. Cup-shaped nest is built on base of palm fronds, among palm seed clusters, or bromeliads, made of dry grasses, rootlets, hair, and feathers. Lays three or four grayish-white eggs, with gray and brown spots concentrated at large end. **Voice**: A highly variable *tew-tew-tew*, or *tee-leeoo* often repeated and also a prolonged, nasal *enk*, a chatter call, and a simple *kik*, often repeated. **Food**: Omnivorous.

RUSTY BLACKBIRD*
TOTÍ PARDO

Euphagus carolinus   Pl. 51

Description. 9" (23 cm). *Male*: blackish, with rusty feather tips only in fall; barred below. Eye very pale yellow, appearing white in the field; eyebrow broad and buffy. Breeding plumage is black with very faint green gloss. Very slender fine-pointed bill. *Female*: in fall, similar to nonbreeding male, with buff underparts. Breeding plumage is slate. Immature similar to winter adults.

Similar species. Greater Antillean Grackle is larger and has keel-shaped tail.

RANGE. Northern North America, wintering to southeastern United States. **Status**: A single sight report: Oct (1960s) La Habana. **Habitat**: Open forest. **Voice**: A low *chuck* and various high squeaky notes. **Food**: Insects, seeds.

GREATER ANTILLEAN GRACKLE

CHICHINGUACO; HACHUELA

Quiscalus niger   Pl. 49

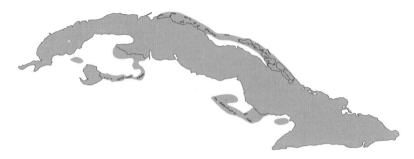

DESCRIPTION. 11" (28 cm). Entirely black, very glossy in male, with metallic blue and violet sheen and bright yellow eye. Tail creased or keel shaped. Female is similar, slightly smaller. Juvenile is similar to adults, with brownish wings and eye. The race *caribaeus* of extreme western Cuba, Isla de Pinos, and adjacent cays has a metallic green gloss. Social, though often quarreling among themselves; hundreds gather at night to roost in groves of tall trees. Often seen bathing in puddles during the heat of the day.

SIMILAR SPECIES. Cuban Blackbird has a shorter and thicker bill, brown eye, and flat tail.

RANGE. Greater Antilles, Cayman Islands. **Status**: Abundant permanent resident in Cuba, Isla de Pinos, and many of the larger cays. **Habitat**: Swamps, coastal vegetation, mangroves, farmland, parks, villages. **Nesting**: Mar–Jul. The cup-shaped nest is built of grasses and mud, usually on base of palm fronds or large bromeliads. Lays three to five olivaceous eggs, with reddish brown spots and scrawls. **Voice**: Calls include a loud, harsh *chin-chin-chin-lin-lin*; also a loud *tik*, often repeated. Adult males display with tail spread, wings drooped, and feathers fluffed out, then produce a loud, metallic *weeee-seeee-eeee*. **Food**: Omnivorous, including *Anolis* lizards and snails.

SHINY COWBIRD

PÁJARO VAQUERO; TORDO LUSTROSO

Molothrus bonariensis   Pl. 48

DESCRIPTION. 7.5" (19 cm). *Male*: very deep metallic violet blue, with greenish wings, usually appearing black overall. *Female*: dark grayish brown above; pale grayish brown below. Both sexes have square, flat tail. Immature similar to female, but with dusky streaks.

SIMILAR SPECIES. (1) Female Brown-headed Cowbird has larger bill and streaked underparts. (2) Tawny-shouldered Blackbird is black, with orange brown shoulder patch.

RANGE. Panama to Southern Argentina; Lesser Antilles. Began invading Greater Antilles from the southeast in 1940s, reaching Cuba in the late 1970s. Also in the Bahamas and southeastern United States. **Status**: Common permanent resident on Cuba and Isla de Pinos. **Habitat**: Farmland, parks, villages, open country. **Nesting**: Mar–Jul. A brood parasite, laying its eggs in the nests of other species. In Cuba, known to parasitize Black-whiskered Vireo and Black-cowled Oriole. **Voice**: Male produces a loud, hurried, musical song characteristically delivered in flight. Commonly heard call is a rapid *tut-tut-tut-tut-tut*. **Food**: Seeds, insects.

BROWN-HEADED COWBIRD

TOTÍ AMERICANO

Molothrus ater   Pl. 48

DESCRIPTION. 7.5" (19 cm). *Male*: black with green and purple metallic sheens and dark brown head. *Female*: grayish brown, somewhat darker above and faintly streaked below. Both sexes have short and square tail. Juvenile is similar to female, conspicuously streaked below. Molting juvenile male has an irregular pattern of black blotches on gray.

SIMILAR SPECIES. Female Shiny Cowbird has smaller bill and is unstreaked below.

RANGE. Breeds in North America, wintering to Mexico. **Status**: Vagrant. Two records: 20 Jan; 12 Feb. Both from Fomento, Villa Clara. **Habitat**: Open woodlands, farmlands. **Voice**: Most characteristic note, by female, is rattled *tut-tut-tut-tut-tut*, almost identical to a call of Shiny Cowbird. In flight, males give a squeaky whistle. **Food**: Insects, seeds.

BLACK-COWLED ORIOLE

SOLIBIO; GUAINUBA

Icterus dominicensis   Pl. 50

DESCRIPTION. 8" (20 cm). Sexes alike. Black with yellow patches on shoulder, underwing coverts, rump, and tail coverts. Immature and subadult are

olivaceous green, with black throat, chest, and forehead, and patches as in adult but yellowish green. Lacks wing bars. Flight is undulating, producing a churring sound with the wings.

SIMILAR SPECIES. First-spring male Orchard Oriole has black markings on back, contrasting with unmarked rump. Female has two greenish-white wing bars.

RANGE. Mexico, Central America, northern Bahamas, and Greater Antilles except Jamaica. **Status**: Common permanent resident in Cuba and Isla de Pinos, and some northern cays such as Guillermo, Coco, and Paredón Grande. **Habitat**: Forests, from sea level to moderate elevations, farmland, parks, villages. **Nesting**: Feb–Jul. Builds elaborate globular nest constructed of palm fibers, sewn to underside of a palm frond, with a side entrance. Lays three greenish-white eggs, with brown and lilac spots concentrated at large end. **Voice**: Song, a series of 8–12 clear, varied down-slurred whistled notes. Call, a harsh *tick*. **Food**: Fruits, nectar, flowers, butterflies; other insects.

ORCHARD ORIOLE
TURPIAL DE HUERTOS

Icterus spurius   Pl. 50

DESCRIPTION. 7.25" (18 cm). *Male*: black head, breast, and upper back; chestnut belly, lower back, rump, and lesser wing coverts. *Female*: olive above; yellowish green below. Wing dark with two whitish wing bars. Both sexes have slightly rounded tails. Immature like female; first-year male has black on chin, throat, and chest; back is mottled with black.

SIMILAR SPECIES. (1) Baltimore Oriole is slightly larger and orange yellow below, with longer bill. (2) Hooded Oriole has longer, more graduated tail. Immature male has black lores; immature female has more decurved bill;. (3) Black-cowled Oriole (immature and subadult) has plain wings, olivaceous back, and yellowish green patches on shoulder, lower back, rump, and undertail coverts.

RANGE. Breeds in southern Canada, eastern United States, and Mexico, wintering to Colombia and Venezuela. **Status**: Rare transient in Cuba; very rare during fall, less so in spring (Oct, 1 Nov; 6 Apr–13 May). One winter record: 20 Jan. **Habitat**: Forests, coastal vegetation. **Voice**: A soft *chuck*. **Food**: Insects, spiders, worms, fruits.

HOODED ORIOLE
TURPIAL DE GARGANTA NEGRA
Icterus cucullatus   Pl. 50

DESCRIPTION. 8" (20 cm). *Male*: yellowish orange, with black upper back, wings, tail, throat, and chest. *Female*: greenish yellow; slightly paler below. Both sexes have long, slightly decurved bill, wing bars, and long and markedly graduated tail. Immature similar to female; male black from lores to upper breast.

SIMILAR SPECIES. (1) Orchard Oriole has shorter, somewhat less graduated tail. Female has yellower underparts. Immature male lacks black on lores, and may have traces of chestnut in spring. (2) Female Northern Oriole has orange yellow underparts.

RANGE. Southwestern United States and Mexico, wintering within the breeding range. **Status**: Vagrant. Two specimens from 19th century, La Habana and Matanzas provinces. **Voice**: A metallic *sheenk*. **Food**: Insects, nectar, fruits.

BALTIMORE ORIOLE
TURPIAL
Icterus galbula   Pl. 50

DESCRIPTION. 8.25" (21 cm). *Male*: black above, with orange lower back, rump, and outer margins of tail. Black head; orange breast and belly. *Female*: crown and upper back mostly brownish orange; underparts dull orange yellow. Immature male similar to adult female, but with black feathers emerging around face and throat; immature female duller.

SIMILAR SPECIES. Nonadult male Orchard Oriole is yellowish green below, with shorter bill. (2) Hooded Oriole is also yellowish green below, with longer and decidedly decurved bill.

RANGE. Breeds in North America and northern Mexico, wintering to Colombia and Venezuela. Greater Antilles. **Status**: Uncommon winter resident and transient in Cuba, Isla de Pinos, and Cayo Coco (15 Sep–16 May). **Habitat**: Open forests, edges of swamps, gardens. **Voice**: A series of rapid, clear, flutelike whistles, though generally silent when wintering. **Food**: Fruits, large insects, nectar.

FINCHES Fringillidae

Small to medium sized birds, often brightly colored, with compact bodies, conical bills, and undulating flight. Typically inhabit forest or open country. Outside the nesting season they are sociable, often travelling in flocks. They feed mostly on seeds and insects. Most species are sexually dimorphic. (**W**:124; **C**:1)

AMERICAN GOLDFINCH*

GORRIÓN AMARILLO

Carduelis tristis   Pl. 51

DESCRIPTION. 5" (13 cm). In winter, males are olive brown above with blackish wings and tail, conspicuous buff-white wing bars, and yellow wash on head, flanks brownish buff. Females duller. In breeding plumage, male is bright yellow with black cap, black wings and tail, thin white wing bars, white rump, and a yellow shoulder patch. Females duller overall, lacking black cap and yellow shoulder patch.

SIMILAR SPECIES. None.

Distribution: Breeds from southern Canada south through northern two-thirds of United States. Winters in southern United States to extreme northern Mexico. **Status**: Six observed: 6 Apr (1978). Cárdenas. **Habitat**: Gardens. **Voice**: In flight a repeated *tip-tip* in synchrony with undulations. **Food**: Seeds.

OLD WORLD SPARROWS Passeridae

These finch-like birds are robust and have relatively large heads, with stout conical bills and short legs. Most are notably gregarious; several build large elaborate colonial nests. They often roost in large flocks. They feed mostly on seeds and insects. Some species are sexually dimorphic. (**W**:37; **C**:1)

HOUSE SPARROW

GORRIÓN; GORRIÓN DOMÉSTICO

Passer domesticus   Pl. 47

DESCRIPTION. 6.25" (16 cm). *Male*: breeding plumage brown above with black streaks and a single conspicuous white wing bar. Gray crown; chestnut nape; black bill. Grayish brown below, with extensive black patch on throat and breast. Nonbreeding plumage duller, with yellow lower mandible and black patch on chest edged with gray. *Female*: grayish brown above with darker streaks; plain grayish brown below. Bill yellowish. Wings brown. Crown brownish gray; eyebrow pale. Juvenile similar to female.

SIMILAR SPECIES. (1) Immature White-crowned Sparrow has two wing bars; head boldly striped with dark brown. (2) Bobolink also has striped head as well as sharply pointed tail feathers and a much warmer tone overall.

RANGE. Eurasia and North Africa. Introduced worldwide. **Status**: Common throughout Cuba and Isla de Pinos. Recorded on Cayo Coco, but not established. **Habitat**: Cities and towns; absent in areas remote from human habitation. **Nesting**: Jan–Aug. Nest is rather large and messy, vaguely cup shaped. Lays three or four white eggs, very variably marked with spots, speckling, or small blotches of gray, greenish gray, black, or brown. **Voice**: Loud *cheep* and *chizzik*. **Food**: Omnivorous.

ESTRILDID FINCHES Estrildidae

Small, mostly terrestrial seed-eating birds of the Old World tropics. All are compact with conical bills; many are brightly colored and prized as cage birds. Unfortunately, escapes from captivity are common, and some species have become established in the wild outside their normal ranges. (**W**:140; **C**:1)

CHESTNUT MANNIKIN
MONJA TRICOLOR
Lonchura malacca   Pl. 47

DESCRIPTION. 4.5″ (11.5 cm). Strikingly marked with black hood, bluish bill, cinnamon back, and a black patch on white belly. Immature brown.
SIMILAR SPECIES. None.
RANGE. India through southeastern Asia to the Philippines. Introduced in Puerto Rico, reaching Cuba by unknown means, possibly hurricanes. **Status**: Widespread but locally common. First specimen collected near Aguada de Pasajeros, in summer 1990. Abundant in rice fields at Zapata peninsula. **Habitat**: Rice fields. **Nesting**: Apr–Aug. A bulky nest with the entrance hole in the side. Known to also nest in sugar cane fields in Puerto Rico. Lays four or five white eggs. **Voice**: A nasal *honk*. **Food**: Seeds, insects.

Appendix 1

ENDEMIC SPECIES LIST

Below is a list of the species found only in Cuba, together with their respective subspecies, where appropriate. Their distribution is described only in a general way.

Scientific Name	Distribution
Gundlach's Hawk	
Accipiter gundlachi gundlachi	Western provinces.
Accipiter gundlachi wileyi	Easternmost mountain ranges.
Zapata Rail	
Cyanolimnas cerverai	Zapata peninsula.
Blue-headed Quail-Dove	
Starnoenas cyanocephala	Widespread.
Cuban Parakeet	
Aratinga euops	Guanahacabibes, Zapata, Casilda, Najasa, Ciego de Ávila province.
Cuban Screech-Owl	
Otus lawrencii	Widespread.
Cuban Pygmy-Owl	
Glaucidium siju siju	Widespread on mainland, except western tip.
Glaucidium siju vittatum	Guanahacabibes and Isla de Pinos.
Bee Hummingbird	
Mellisuga helenae	Guanahacabibes, Zapata, montane east.
Cuban Trogon	
Priotelus temnurus temnurus	Widespread on main island; Cayo Guajaba and Cayo Sabinal.
Priotelus temnurus vescus	Isla de Pinos.
Cuban Tody	
Todus multicolor	Widespread.

Cuban Green Woodpecker

Xiphidiopicus percussus percussus — Widespread on main island; Lanzanillo, Conuco, Santa María, Coco, and other cays.

Xiphidiopicus percussus insulaepinorum — Isla de Pinos.

Xiphidiopicus percussus monticola — Mountain ranges east of Baracoa.

Xiphidiopicus percussus gloriae — Cayo Cantiles.

Xiphidiopicus percussus cocoensis — Cayo Guillermo.

Fernandina's Flicker

Colaptes fernandinae — Soroa, Zapata peninsula, Najasa.

Zapata Wren

Ferminia cerverai — Zapata peninsula.

Cuban Gnatcatcher

Polioptila lembeyei — Cayo Coco, Cabo Cruz to Maisí, Nuevitas.

Cuban Solitaire

Myadestes elisabeth — Sierra del Rosario and eastern mountain ranges.

Cuban Vireo

Vireo gundlachii gundlachii — West from Camagüey province; Isla de Pinos.

Vireo gundlachii orientalis — Southern coast of easternmost provinces, Santa María, Cayo Guillermo and Cayo Coco.

Vireo gundlachii magnus — Cayo Cantiles.

Vireo gundlachii sanfelipensis — San Felipe cays.

Yellow-headed Warbler

Teretistris fernandinae — Western Cuba.

Oriente Warbler

Teretistris fornsi — Eastern Cuba.

Red-shouldered Blackbird

Agelaius assimilis — Western and central Cuba.

Cuban Blackbird

Dives atroviolacea — Widespread.

Cuban Grassquit

Tiaris canora — Widespread.

Zapata Sparrow

Torreornis inexpectata inexpectata — Santo Tomás, Zapata peninsula.

Torreornis inexpectata sigmani — Baitiquirí, Guantánamo province.

Torreornis inexpectata varonai — Cayo Coco.

ENDEMIC SUBSPECIES

In addition, below we provide a list of the subspecies or races exclusive to Cuba, with a brief range description, so that they can be noted separately from the North American forms. Only a general description of their distribution is presented.

Scientific Name	Distribution
Chondrohierax uncinatus wilsonii	Sagua-Baracoa mountain range.
Accipiter striatus fringilloides	Local, mountains.
Buteo platypterus cubanensis	Widespread.
Buteogallus anthracinus gundlachii	Widespread on coasts.
Falco sparverius sparveroides	Widespread.
Colinus virginianus cubanensis	Widespread.
Grus canadensis nesiotes	Viñales valley, Pinar del Río; Isla de Pinos; north of Villa Clara, Ciego de Ávila, and Camagüey province; Zapata peninsula.
Rallus elegans ramsdeni	Widespread.
Geotrygon caniceps caniceps	Local; common in Zapata peninsula.
Amazona leucocephala leucocephala	Guanahacabibes; Mil Cumbres; Zapata peninsula; Isla de Pinos; Najasa; Escambray, eastern Cuba, Ciego de Ávila province.
Saurothera merlini merlini	Widespread on main island; Cayo Conuco, Caibarién.
Saurothera merlini decolor	Isla de Pinos.
Saurothera merlini santamariae	Santa María, Coco, Paredón Grande, and Romano cays.
Asio stygius siguapa	Zapata peninsula; Escambray; Isla de Pinos: La Habana.
Caprimulgus cubanensis cubanensis	Widespread.
Caprimulgus cubanensis insulaepinorum	Isla de Pinos.
Tachornis phoenicobia iradii	Local, main island.
Chlorostilbon ricordii ricordii	Widespread.
Melanerpes superciliaris superciliaris	Main island and Cayo Cantiles.
Melanerpes superciliaris murceus	Isla de Pinos.
Melanerpes superciliaris florentinoi	Cayo Largo.
Melanerpes superciliaris sanfelipensis	Cayo Real (San Felipe cays).
Colaptes auratus chrysocaulosus	Main island; Coco, Paredón Grande, and Romano cays.
Campephilus principalis bairdii	Mountains south of Moa (possibly extinct).
Tyrannus caudifasciatus caudifasciatus	Widespread on main island, Isla de Pinos; several cays.
Contopus caribaeus caribaeus	Main island; Isla de Pinos; Coco and Paredón Grande cays.
Contopus caribaeus morenoi	Zapata peninsula and Archipiélago de los Canarreos.

Contopus caribaeus nerlyi	Archipiélago de los Jardines de la Reina.
Contopus caribaeus florentinoi	Archipiélago de los Jardines de la Reina.
Contopus caribaeus sanfelipensis	San Felipe cays.
Petrochelidon fulva cavicola	Main island; Isla de Pinos.
Spindalis zena pretrei	Main island; Isla de Pinos; Cantiles, Francés, Santa María, Coco, Paredón Grande, and Romano cays.
Melopyrrha nigra nigra	Main island; Isla de Pinos; Cantiles, Guillermo, Coco, Romano, and other cays.
Agelaius humeralis humeralis	Widespread on main island; Isla de Pinos; Cayo Coco; Jardines de la Reina.
Agelaius humeralis scopulus	Cayo Cantiles.
Sturnella magna hippocrepis	Widespread on main island; Isla de Pinos; Cayo Romano.
Quiscalus niger gundlachii	Widespread on main island, except western tip; Coco, Romano, Guajaba, and other cays.
Quiscalus niger caribaeus	Western tip of main island; Isla de Pinos; southern and some northern cays.
Icterus dominicensis melanopsis	Main island, Isla de Pinos, and several cays.

Appendix 2

GLOSSARY

Arboreal Frequents trees.

Auricular See *Earspot*.

Axillaries The long, innermost feathers of the underwing, covering the area where the wing joins the body.

Back The portion of the upperparts located behind the nape and between the wings.

Bar A long marking across the body or across a feather; a transverse marking (see *Stripe*).

Belly The portion of the underparts between the breast and the undertail coverts.

Bend of wing The most anterior part of the folded wing.

Breast The area of the underparts between the foreneck and the belly.

Breeding (nuptial) plumage The more colorful plumage acquired by adults of some bird species immediately before the reproductive period. In the nonbreeding season it is replaced by a more subdued plumage.

Bristle A specialized feather usually found at the junction of the upper and lower mandibles that may amplify the sensations of touch or may help to filter out wood or other particles; also may protect the eye.

Call The short unmusical vocalizations emitted year-round by both sexes. These are characteristic for each species and given in many different social and feeding situations, including flushing, alarm, etc.

Cap An area of distinct color on the top of the head that may reach the nape.

Carnivorous Animal-eating.

Cheek The side of the face.

Chest See *Breast*.

Chin The area immediately below the base of the lower mandible.

Clutch A complete set of eggs laid by an individual female during a single nesting.

Collar A band of distinct color partially or wholly encircling the neck.

Colonial Nesting in groups (or colonies) rather than in isolated pairs.

Color morph Marked individual differences in color pattern within a population not related to age, sex, or season.

Coronal stripe A stripe of contrasting color along the center of the crown.

Coverts Small feathers that cover the bases of the large flight feathers of the wings and tail, or that cover a particular area or structure (e.g., ear coverts).

Crepuscular Active at twilight.

Crest Elongated feathers on top of the head.

Crown The top of the head.

Crown patch A contrasting color patch on top of the head.

Cryptic Serving to conceal.

Dichromatic Having two color phases, not related to age or sex.

Dimorphic Having two distinct forms within a population, differing in size, form, or color, often related to sex.

Diurnal Active during the day.

Earspot A contrasting color mark on the ear area.

Ear tuft Long feathers arising from the top of the head, suggesting ears, a characteristic of certain owls.

Eclipse Postnuptial plumage stage occurring in waterfowl wherein males become duller, resembling females, and are flightless for a brief period.

Endemic A species occurring in only one country or geographical area.

Exotic A species not native to a given geographical area.

Extinct A species with no living representatives.

Extirpated A species no longer occurring in a given geographical area, usually as a result of human interference, but still surviving elsewhere.

Eyebrow A stripe on the side of the head immediately above the eye; also known as *supercilium*.

Eyeline A straight, thin, horizontal stripe on the side of the face, through the eye.

Eye ring A fleshy or feathered ring around the eye, often distinctively colored.

Eye stripe A stripe that runs horizontally from the base of the bill through the eye; usually broader than an eyeline.

Facial disk A peculiar pattern of feathers that encircles the eyes of some birds, especially the owls.

Facial markings Contrasting color markings on the side of the head.

Feral Wild; here applies to domestic animals living in the wild.

Field mark A plainly visible characteristic, such as color, shape, pattern, or behavior, useful in identifying a bird species in the field.

Flank The rear portion of the side of a bird's body.

Forehead The area of the head just above the base of the upper mandible.

Foreneck The front or underside of the neck.

Fossil bird Any member of the class Aves, living or extinct, whose remains have been preserved in stone caverns or beneath dried-up and former lakes.

Frontal shield A hard, featherless plate extending from the base of the upper mandible over the forehead; typical of coots and gallinules.

Frugivorous Fruit-eating.

Gape The angle between the upper and lower mandibles when the bill is open.

Genus Taxonomic group of closely related species; plural is *genera*.

Gleaning A feeding technique in which insect prey are picked from the surface of leaves and small branches.

Gliding Usually brief flight supported by motionless, extended wings.

Greater Antilles The larger Caribbean islands: Cuba, Hispaniola, Jamaica, and Puerto Rico.

Gregarious Forming flocks.

Hindcrown The rear portion of the crown.

Hindneck The rear or upper surface of the neck; the nape.

Hood A well-defined color area covering most or all of the head.

Hovering The most energy-consuming form of flight that allows some birds to maintain the same position in midair through constant wing flapping.

Hybrid An individual resulting from the interbreeding of two different species.

Immature plumage The sometimes differently colored plumage(s) of young birds before they reach sexual maturity.

Insectivorous Insect-eating.

Introduced species An exotic one, accidentally or purposely brought from abroad.

Juvenile A bird in juvenile plumage.

Juvenile plumage The first plumage that replaces natal down, which includes true contour feathers; usually replaced, by immature or adult plumage, in late summer or fall.

Lesser Antilles The chain of smaller Caribbean islands from the Virgin Islands to Grenada.

Local Occurring in relatively small, restricted areas within the range, rather than widespread throughout the range. Birds with local distributions are often dependent on an uncommon habitat.

Lore The area between the eye and the base of the bill, sometimes distinctively colored.

Mandible One of the two parts of a bird's bill, termed the *upper mandible* and the *lower mandible*.

Mantle The plumage of the back and upper surface of the wings when these areas are of one color; often used in descriptions of gulls.

Mask An area of contrasting color surrounding the eyes.

Middle America All of Central America, south to the Darién Gap, and Mexico, north to the border with the United States.

Migration The regular, two-way movement of bird populations between geographical areas.

Mimic Relating to species that can incorporate sounds of other species in their repertoire.

Molt A periodical and gradual process in which birds partially or totally replace old feathers with new ones.

Mustache A broad colored streak running from the base of the bill back along the side of the throat.

Nape That part of the hindneck just below the base of the skull.

Native A species indigenous to a specific geographical area; one naturally breeding there.

Necklace A band of spots or streaks across the breast or around the neck.

Nocturnal Active at night.

Nostril External opening on the upper mandible.

Nuptial See *Breeding (nuptial) plumage.*

Omnivorous Eating both animal and plant food.

Pellet Indigestible portions of food regurgitated by various bird species, particularly owls.

Permanent resident A native species that spends the whole year in one geographical area.

Plumes Long ornamental feathers.

Postnuptial See *Winter (basic) plumage.*

Postocular stripe A stripe extending back from the eye, above the ear coverts and below the eyebrow.

Primaries The outermost and longest flight feathers on a bird's wing. There are usually 9–12 primaries per wing.

Probing A feeding technique that involves introducing the bill into natural cavities, flowers, mud, or sand, in search of food items.

Range The geographical area or areas normally inhabited by a species.

Rump The lower back just above the tail; may also include the uppertail coverts.

Scapulars The feathers of the upperparts at the side of the back that cover the area where the wing joins the body.

Secondaries The large flight feathers of the inner wing, attached to the inner wing proximal to the wrist.

Sexual dimorphism A difference between the sexes in size, form, or color.

Shaft The stiff central axis of a feather; also known as the *rachis.*

Side The lateral part of the breast and belly.

Soaring flight The less energy-consuming form of flight in which a bird uses rising air masses (thermals) or currents to gain height with wings extended and little or no flapping.

Song An often complex and extended vocalization generally used in territorial advertisement and disputes, and for communication with mates; more commonly used by males during the breeding season.

Species A group of actually or potentially interbreeding populations reproductively isolated from other such groups.

Spectacles A color pattern formed by the lores and eye rings.

Speculum A patch on the wing, usually rectangular, contrasting in color with the rest of the wing and often brightly colored; typical of many species of waterfowl.

Stripe Elongated marking running lengthwise on the bird (see *Bar*).

Subadult A bird that has not attained definitive plumage typical of the adult.

Subspecies or geographical races Local, and often isolated, populations of a species that exhibit a different combination of external characters, such as color, size, or in some cases, vocalizations.

Subterminal Before, or short of, the end or tip.

Supercilium See *Eyebrow.*

Tail streamer Slender, elongated tail feathers.

Taxonomy The science of naming and classifying animals and plants, organized in hierarchies according to perceived evolutionary relationships among the different species, or other groups, such as families or genera.

Territorial A species that defends a territory (see *Territory*).

Territory An area defended by the male, by both members of a pair, or by an unmated bird, often for the purposes of nesting or defending food supplies.

Tertials The innermost secondaries on a bird's wing, immediately adjacent to the body.

Thermal A vortex of rising warm air produced by solar heating of the ground.

Threatened Restricted populations vulnerable to various pressures.

Throat The area of the underparts between the chin and the breast.

Throat pouch An area of bare skin on the throat that may be brightly colored and sometimes inflatable; also known as *gular sac*.

Underparts The lower surface of the body, including the chin, throat, breast, belly, sides and flanks, and undertail coverts, and sometimes including the underwing surface and the under surface of the tail.

Undertail coverts The small feathers that lie beneath, and cover the base of, the tail feathers.

Upperparts The upper surface of the body, including the crown, nape, back, scapulars, rump, and uppertail coverts, and sometimes including the upperwing surface and the upper surface of the tail.

Uppertail coverts The small feathers that lie above, and cover the base of, the tail feathers.

Vane One of the two broad, thin flexible portions of a feather, separated by the shaft and composed of a row of barbs that are connected along the shaft; also called a *web*.

West Indies Includes the Bahama Islands, Greater Antilles, the Cayman Islands, Swan Islands, Old Providence and St. Andrew islands in the southwestern part of the Caribbean, and the Lesser Antilles south to Barbados and Grenada.

Whisker stripe A narrow colored streak running from the base of the bill back along the side of the throat (see *Mustache*).

Wing bar A crosswise stripe on the folded wing, formed by the tips of the wing coverts.

Wing lining A collective term for the coverts of the underwing.

Wing stripe A conspicuous stripe running the length of the open wing.

Winter (basic) plumage The generally duller plumage acquired by some bird species after breeding, and worn throughout the winter.

Wrist The forward-projecting angle or bend of the wing; also called the *carpal joint*.

References

Academia de Ciencias de Cuba and Instituto Cubano de Geodesia y Cartografía. 1989. *Nuevo Atlas Nacional de Cuba*. Rhea e Instituto Geográfico Nacional, Spain.

American Ornithologists' Union. 1983. *Check-list of North American Birds* (6th ed.). Allen Press, Lawrence, Kansas. 877 pp.

——1989. Thirty-seventh supplement to the AOU Check-list of North American Birds. *Auk* 106: 532–538.

——1991. Thirty-eighth supplement to the AOU Check-list of North American Birds. *Auk* 108: 750–754.

——1993. Thirty-ninth supplement to the AOU Check-list of North American Birds. *Auk* 110: 675–682.

——1995. Fortieth supplement to the AOU Check-list of North American Birds. *Auk* 112: 819–830.

——1997. Forty-first supplement to the AOU Check-list of North American Birds. *Auk* 114: 542–552.

——1998. Check-list of North American Birds (7th ed.). Allen Press, Lawrence, Kansas. 829 pp.

Arredondo, O. 1984. Sinopsis de las aves halladas en depósitos fosilíferos Pleistoceno-Holoceno de Cuba. Academia de Ciencias de Cuba. *Reporte de Investigación* 17: 35.

Bond, J. 1948. Origin of bird fauna of the West Indies. *Wilson Bulletin* 60: 207–229.

——1968. Derivation of the Antillean avifauna. *Proceedings of the Academy of Natural Sciences of Philadelphia* 115(4): 79–98.

——1985. *Birds of the West Indies* (5th ed.). Houghton Mifflin, Boston. 256 pp.

Borhidi, A. 1983. The main vegetation units of Cuba. *Acta Botanica Hungarica* 33(3–4): 151–185.

Bradley, P. 1995. *Birds of the Cayman Islands*. Caerulea Press, Italy. 261 pp.

Capote, R. P. and R. Berazaín. 1984. Clasificación de las formaciones vegetales de Cuba. *Revista Jardín Botánico Nacional*. 5(2): 27–75.

Curson, J., D. Quinn, and D. Beadle. 1994. *New World Warblers*. Helm Identification Guides, London. 252 pp.

Ehrlich, P. R., D. S. Dobking, and D. Wheye. 1988. *The Birder's Hand Book. A Field Guide to the Natural History of North American Birds*. Simon and Schuster, New York. 785 pp.

Farrand, J. (ed.). 1989. *The Audubon Society Master Guide to Birding*. Vols. 1–3. Alfred A. Knopf, New York. Vol 1: 447 pp.; Vol 2: 398 pp.; Vol 3: 399 pp.

Garrido, O. H. and F. Garcia Montaña. 1975. *Catálogo de las aves de Cuba*. Academia de Ciencias de Cuba, La Habana. 149 pp.

Garrido, O. H. and A. Kirkconnell (n.d.). Catalogue of Cuban birds. Unpublished manuscript.

Godfrey, W. E. 1966. The Birds of Canada. *National Museum of Canada Bulletin* 203, *Biology Series* 73. Ottawa, Ontario. 428 pp.

Gundlach, J. 1876. *Contribución a la Ornitología Cubana*. Imprenta La Antilla, La Habana. 364 pp.

Harrison, C. 1987. *A Field Guide to the Nests, Eggs and Nestlings of North American Birds.* Collins, London. 416 pp.

Harrison, P. 1983. *Seabirds.* Croom Helm, London. 448 pp.

Hayman, P., J. Marchant, and T. Prater. 1986. *Shorebirds. An Identification Guide.* Houghton Mifflin, Boston. 412 pp.

Howard, R. and A. Moore. 1991. *A Complete Checklist of the Birds of the World.* Academic Press, New York. 622 pp.

Instituto Cubano de Geodesia y Cartografía. 1978. *Atlas de Cuba.* Academia de Ciencias de Cuba, La Habana.

Iturralde-Vinent, M. A. 1982. Aspectos geológicos de la biogeografía de Cuba. *Revista Ciencia Tierra y Espacio.* 5: 55–100.

——1994. Cuban geology: A new plate tectonic synthesis. *Journal of Petroleum Geology* 17:39–70.

International Union for Conservation of Nature (IUCN). 1996. *1996 IUCN Red List of Threatened Animals.* IUCN, Gland, Switzerland. 368 pp.

Kaufman, K. 1990. *Advanced Birding.* Houghton Mifflin, Boston. 299 pp.

Kirkconnell, A., O. H. Garrido, R. M. Posada, and S. Cubillas. 1992. Los grupos tróficos en la avifauna Cubana. Academia de Ciencias de Cuba. *Poeyana* 415: 21.

MacPhee, R. D. E. and M. A. Iturralde-Vinent. 1994. First tertiary land mammal from Greater Antilles: an early miocene sloth (*Xenarthra*, Megalonychidae) from Cuba. *American Museum Novitates* 3094: 1–13.

——1995. Origin of the Greater Antillean land mammal fauna, 1: New tertiary fossils from Cuba and Puerto Rico. *American Museum Novitates* 3141: 1–30.

National Geographic Society. 1987. *Field Guide to the Birds of North America.* National Geographic Society, Washington, D.C. 464 pp.

Peterson, R. T. 1980. *A Guide to Birds East of the Rockies.* Houghton Mifflin, Boston. 384 pp.

Pyle P., S. N. G. Howell, R. P. Yunick, and D. F. Desante. 1987. *Identification Guide to North American Passerines.* Slate Creek Press, Bolinas, California. 278 pp.

Ricklefs, R. E. and E. Bermingham. 1997. Molecular phylogenetics and conservation of Caribbean birds. *El Pitirre* 3(10): 85–92.

Sibley, G. and B. L. Monroe. 1990. *Distribution and Taxonomy of Birds of the World.* Yale University Press, New Haven, Connecticut. 1111 pp.

Suárez, W. and O. Arredondo. 1997. Nuevas adiciones a la paleornitología Cubana. *El Pitirre* 3(10): 100–102.

Veit, R. and L. Jonsson. 1987. Field identification of smaller sandpipers within the genus *Calidris. American Birds* 41: 213–236.

CONSULTED COLLECTIONS

American Museum of Natural History, New York.
Academy of Natural Sciences of Philadelphia, Philadelphia.
Museum of Comparative Zoology, Harvard University, Boston.
Lousiana State University Museum of Natural Science, Baton Rouge.
Museo Nacional de Historia Natural de Cuba, La Habana.
Instituto de Ecología y Sistemática, La Habana.
United States National Museum, Smithsonian Institution, Washington, D.C.

Index

English common names appear in **bold**.
Technical names appear in *italics*. Local Cuban names are in normal text.
Numbers in **bold** specify the plate number.